The Dark Side of the Hive

The Dark Side of the Hive

THE EVOLUTION OF THE IMPERFECT HONEY BEE

Robin Moritz and Robin Crewe

OXFORD
UNIVERSITY PRESS

OXFORD
UNIVERSITY PRESS

Oxford University Press is a department of the University of Oxford. It furthers the University's objective of excellence in research, scholarship, and education by publishing worldwide. Oxford is a registered trade mark of Oxford University Press in the UK and certain other countries.

Published in the United States of America by Oxford University Press
198 Madison Avenue, New York, NY 10016, United States of America.

Library of Congress Cataloging-in-Publication Data
Names: Moritz, Robin F. A., and Crewe, Robin M.; authors.
Title: The dark side of the hive / Robin Moritz and Robin Crewe.
Description: New York, NY : Oxford University Press, 2018. |
Includes bibliographical references and index.
Identifiers: LCCN 2018010888 | ISBN 9780190872281
Subjects: LCSH: Honeybee—Life cycles.
Classification: LCC QL568.A6 M594 2018 | DDC 595.79/9156—dc23
LC record available at https://lccn.loc.gov/2018010888

{ CONTENTS }

{ PREFACE }

Producing a monograph on the life history of the honey bee, *Apis mellifera*, may seem somewhat quixotic at first sight in view of the fact that there are a series of excellent monographs on this topic that have been published during the past three decades. However, we believe that these volumes have a bias that is too strongly focused on the harmony and the perfection of cooperation in the colony. Tom Seeley's excellent treatises *Honeybee Ecology* (1985), *The Wisdom of the Hive* (1995), and, most recently, *Honeybee Democracy* (2010) have had a major influence on students of honey bee biology. Indeed, they stand out as seminal works in organismal biology in the mold of the work of von Frisch, who won the Nobel Prize for his work on honey bees. Mark Winston's book, *The Biology of the Honey Bee* (1987), as well as Robin Moritz and Edward E. Southwick's *Bees as Superorganisms—An Evolutionary Reality* (1991) have dealt in great detail with the marvels of cooperation inside the colony. Yet as with any complex social system, honey bee societies are prone to error, robbery, cheating, and social parasitism. The honey bee colony is thus far from being a harmonious, cooperative whole. It is full of individual mistakes, obvious maladaptations, and evolutionary dead ends. Conflict, cheating, worker inefficiency, and curious reproduction strategies all occur. The perfection that is perceived to exist in their social organization is a function of a particular experimental focus on the colony as a whole rather than exploring the idiosyncrasies of its individual members.

The fact that honey bee colonies get by remarkably well despite many seemingly odd biological features that are often dismissed as aberrations requires us to focus attention on these very "aberrations" because they are central to understanding all aspects of social organization. Since we have worked together for more than two decades on the chemical ecology, genetics, and evolution of parasitic honey bee workers, we believed it is now overdue to report on the plasticity of social organization in the honey bee colony with a view to achieving a more nuanced understand of the evolution of sociality in these insects.

What we cover in this volume is not designed to suggest that the work of our colleagues and our own previous work require revision or reconsideration but, rather, provides a richer understanding of the real life of a honey bee in the colony. Our work thus focuses on the role of the individual within the colony rather than studying the colony as a biological entity (superorganism). We try to dissect the various careers a male and a female honey bee can have and their roles in colony organization.

Colonies of honey bees are made up of a number of families because the queens are multiply mated by a number of drones (males). In order to understand the effects of this genetic diversity within the colony, we provide a thorough discussion of how the variance of individual worker phenotypes drives self-organized processes in a colony.

In addition, we show how competition using both physical force and chemical signaling drives colonial organization. Here, we deal with handicaps that limit the use of physical force and the chemical arms races that drive competition in relation to worker reproduction.

This monograph focuses on bees as individuals in the colony. The story that we tell spans the full range of biological disciplines ranging from genomics to systems biology. We explore the situations in which individual interests are pursued often at the expense of the colony, and we show that the solutions that have evolved are often less than optimal.

{ ACKNOWLEDGMENTS }

This volume would not have been possible without Anne and Mary allowing us to drone on endlessly about honey bees when indeed we should have been much more polite and addressed other more important topics. We are deeply indebted for this indulgence and have no really convincing ideas about how to repay it, particularly because we are not certain that this may not happen again despite solemn undertakings.

We are also grateful to the late Ingemar Fries, Christian Pirk, Anja Buttstedt, and two anonymous reviewers for providing many important comments that markedly improved the text. We considered most of those, but not all; therefore, any errors and misconceptions remain ours and cannot be blamed on others.

The generous funding of many different public institutions and funding organizations that in turn received their money from the taxpayers in Germany, South Africa, and also all European Union member states allowed us to study honey bee biology over several decades. It was this long period of familiarity with honey bee biology that made us bold enough to face the challenge of writing this volume.

During the actual writing phase, this project was significantly facilitated by the receipt of the Harry Oppenheimer Fellowship Award from the Oppenheimer Memorial Trust that allowed RMC to take periods of extended leave in Germany each year from 2014 to 2017. This leave allowed us to work together on the preparation of the manuscript and also to encourage doctoral students to pursue new lines of research.

{ LIST OF IMAGE CREDITS }

Chapter 2

2.1 Redrawn after a photograph provided by N. Koeniger. 8
2.2 Reproduced with permission of the artist Michael Rothman © M. Rothman 1999/2017 9
2.3 Drawings by author (not to scale) based on the analysis of Romiguier et al. (2016). 11
2.4 Drawn by the authors. 13
2.5 Climatic zones on the continents during the Oligocene indicating the benign climates at northern latitudes when honey bee species were becoming widespread. Boucot, A. J., Chen Xu, & Scotese, C. R. (2013). Phanerozoic Paleoclimate: An Atlas of Lithologic Indicators of Climate, SEPM Concepts in Sedimentology and Paleontology, (Print-on-Demand Version), No. 11, 478 p, ISBN 978-1-56576-289-3, October 2013, Society for Sedimentary Geology, Tulsa, OK. 15
2.6 Data used for this figure from Wallberg et al. 2014; Fu et al. 2016. 17

Chapter 3

3.1 Redrawn from Sammataro D, Cicero JM (2010). Functional morphology of the honey stomach wall of the European honey bee (Hymenoptera: Apidae). *Ann Entomol Soc Am* 103: 979–987. 20
3.2 Drawn by the authors. 22
3.3 Drawn by the authors. 23

Chapter 4

4.1 Redrawn after a photograph provided by N. Koeniger. 29
4.2 Reproduced with permission of the authors and republished with permission of the Royal Society, from Peeters CP, Monnin T, Malosse C (1999). Cuticular hydrocarbons correlated with reproductive status in a queenless ant. *Proc Roy Soc B Biol Sci* 266: 1323–1327; permission conveyed through Copyright Clearance Center, Inc. 30

Chapter 5

Chapter 6

Chapter 7

Chapter 8

Chapter 9

Chapter 10

Chapter 11

The Dark Side of the Hive

{ 1 }

Introduction

Of all insecta the Bees are chiefe, and worthily to be most admired; being the only things of that kinde, which are bred for the behoof of men. (Preface)

Among all the Creatures which our bountifull God hath made for the use and service of man, in respect of great profit with smal cost, of their ubiquitie or being in all Countries, and of their continuall labour and comly order, the Bees are most to be admired.

For the Bees abhorre as well Polyarchie, as Anarchie, God having shewed in them unto men, an cxpresse patterne of A Perfect Monarchic, The Most Natural And Absolute Forme Of Government.

—Charles Butler (1609)

1.1. Honey bees and humans

From the caves of Valencia in Spain to those of the southern Drakensberg in South Africa, cave paintings have documented the ancient relationship between honey bees and humans. Honey bees as a source of nourishment were available both to the ancestors of humans and to early human populations because they occurred in the same habitats. Reflection on honey bee social organization has been a source of wonder and of the construction of myths since the origin of hunter–gatherer societies. The interaction between honey bees and humans is of ancient origins both as a source of food via honey and other bee products and as a source of spiritual nourishment through observing their complex social organization that could be used as an exemplar to guide the formation of human societies.

Honey bees are clearly the most studied and best understood social insects on Earth. Over thousands of years, humans have tried to grasp the mystery of how a colony of thousands of small flying insects can possibly exist and manage its relationships. Recording of beekeeping activities dates back to the Egyptian fifth dynasty (2500 BCE), and published honey bee research dates back to Aristotle (350 BCE). Since then, the honey bee has provided the proverbial magic well for research that Karl von Frisch (1965) described and from which researchers of a variety of disciplines have continuously drawn insights and, if they were lucky, even some wisdom. Students of honey bees are attracted to them because their complex social organization seems to be so well ordered: There is a single queen, there are many workers, and together they work for the good of the colony. Obviously, the honey bees manage through marvelous coordination, social intelligence, and sometimes even democratic voting to achieve a level of social harmony that seemingly outperforms human societies. Indeed, honey bee societies are extremely successful and efficient.

The picture of harmony and success is compelling, sometimes perhaps so compelling that it might easily preclude asking critical questions about such obvious efficiency. Yet it is here that the skeptical student of honey bee biology should be wary. Biologists have a strong inclination to compare human societies and bee colonies, by analogy, with a tendency to suggest that the comparison is unfavorable to human societies. The problem with using analogies is that they may misdirect our appreciation of the very core of honey bee biology. Simply by calling the reproductive female "queen" and the sterile females "workers," we have betrayed ourselves through the use of terms that stem from human social systems. Use of these terms implies an implicit conception of social structure that defines the functioning of these animals, which may easily misdirect us from thinking objectively. Of course, queens in human societies have completely different positions and provide absolutely no useful insight into the function of the only female sexual in a colony of many thousand sterile females.

This common modern conception in itself was a foreign one to older male-dominated human societies; hence, in Egyptian hieroglyphics, the honey bee "queen" was considered to be a symbol of the pharaohs and therefore male. However, once one understands that the honey bee "king" of the pharaohs is a queen, then the search for the honey bee "king" becomes an intensive one. By analogy with states that had to be ruled by kings and queens, so had the honey bee colony. It was only when we found out more about the life history and the mating system of the honey bee that we started to understand that social organization in the colony might function in a fundamentally different way from human kingdoms, and through this insight we are freed from some important preconceptions about honey bee social organization, but most likely not all.

1.2. For the good of the society

Today, we believe we understand the functioning of the colony in some detail. We know that the colony is a biologically composite unit that, on the one hand, functions as a single biological entity but, on the other hand, is characterized by the behavior of the many individual organisms within it. This biological structure has repeatedly been conceptualized as a *superorganism*, a term introduced by Wheeler (1911) to help students of social insects grasp the biological peculiarities of the colony and consider the colony as an important unit of selection.

For many years, this has sparked controversies among evolutionary biologists, particularly after the dogma had become established that evolutionary processes had to be exclusively explained at the level of the gene. A student in biology would not have passed an examination in the "selfish gene" era if he or she had invoked group or colony-level selection when explaining evolution in insect social systems. The superorganism was considered to be a misguided conception; group selection was considered to be a public offense and an insult to common sense. Today, the field is more relaxed and the controversy has faded (although not completely), but it is now clear that selection operates on phenotypes including all relevant biological levels: from a nucleotide in a codon within a gene up to the colony level and probably even beyond. Some would like to rekindle the debate by moving to the other extreme of claiming that kin selection is not required for the evolution of sociality at all (Nowak, Tarnita, and Wilson 2010). Many models may explain the evolution of sociality without invoking arguments based on relatedness. However, the transfer of DNA from one generation to the next is the very basis of life in any form. If we ignore this fact in evolutionary philosophy and thinking, we run a great risk of being misled. Ignoring relationships in evolutionary theory is like disregarding gravity in physics. Relatedness and inclusive fitness are inescapable properties of any form of life. It is therefore not only prudent but also essential to include the probability of gene identity in any evolutionary scenario and certainly so in theoretical considerations of the evolution of sociality.

In this monograph, we do not deal with the evolution of sociality. Although the magic well of honey bee research has great depth, honey bees are not a good model system for empirically testing evolutionary theories of the origins of sociality. The entire genus *Apis* includes a suite of species, and all are extremely derived and highly eusocial. Indeed, the oldest fossil *Apis* species found in deposits from the early Oligocene approximately 30 million years ago were already fully eusocial. The specimen is that of a sterile worker. Because the common phylogenetic ancestors of the Apini were already highly eusocial, honey bee social organization is simply not informative for tracing the evolutionary steps from solitary to social life, and it is impossible to determine how honey bees became social if we only search within the genus. However, the honey bee does provide a superb system for studying how natural selection and evolution operate within a society. The colony provides an ideal model to test who wins and loses in the never-ending game of gene transmission from one generation to the next.

1.3. For the good of the individual

By looking into the colony, we can focus on selected individual colony members and groups of individuals and determine whether the evolutionary game of reproduction is a fair one. Efficient cooperation among colony members is clearly the basis for the success of the colony, and there are many rules and mechanisms in place to prevent conflict in the colony. But are the rules always immutable? Does everybody play by the same rules? As with human societies, is there a dark side that we need to explore and understand?

At first sight, the queen seems to go to great lengths to be the winner and to be the major beneficiary, which seems particularly unfair in Darwinian terms. But is this really so? Isn't she rather the egg-laying machine of the workers? These contrasting views define our topic: What may look obvious from one perspective may actually not be the case. What seems plausible at first sight may turn out to be completely wrong when reconsidered. We should remember the long misconception of a honey bee king ruling the colony formalized by Aristotle (350 BCE) because he considered the sting to be a sign of the male sex and the stingless drones to be the females. For more than 1,800 years, it was common sense to have a king in the honey bee colony because who else could rule such a large community? Only after the work of Luis Mendez de Torres (1586) and propagated in English by Charles Butler (1609) did it become clear that the king was rather busy laying eggs and hence might be a female. This was finally confirmed in 1670 when Jan Swammerdam showed that the queen actually had ovaries and was the only fertile female in the colony (Swammerdam 1670/1737). We now know that the concept of the king was utterly wrong, but the identification of a special individual among all the bees with a definitive role was in fact an essential step in reaching today's state of knowledge of honey bee biology.

When reading this volume, we hope that you will join us as we dissect the role of natural selection in shaping the behavior of the individuals that comprise a colony of honey bees—a modest aim that will explore the fierce (and not always fair) controversies in the field.

An understanding of natural selection in honey bees may seem easy from the perspective of the queen because she is unique among the females. The classical Darwinian fitness rules of survival of the fittest apply. However, it is much more difficult to assess natural selection operating on the workers because there are so many of them, they are all different although they look uniform, and most of the time they do not even reproduce. We also try to address the males, which is a challenge not because they are particularly complicated but simply because they have been so little studied since most students of honey bees focus on the female sex and see no need to study the lazy drones. And, yes, this is a gender issue in its truest sense because drones are anything but lazy, and the haploid males are exposed to the effects of natural selection in a fashion that is particularly cruel, as we discuss later.

1.4. For the future

As we embark, we already know that even though we will try to provide the reader with a scientifically solid account, we will very likely generate at least some false conclusions. We cannot exclude making a blunder similar to that of sending the reader off to "search for the honey bee king" simply because we used the wrong evidence or combine ideas that should not have been combined in the first place. We will not argue that we are the only ones to have found the biological truth, but we attempt to draw an up-to-date picture of what individuals in a honey bee colony do to get by in their lives. If the reader believes our arguments and interpretations provide insights for other social systems including that of humans, then we admit that we do not want our narrative to exclude the emergence of such possibilities. In fact, sometimes it is easier to comprehend social behavior in animal systems if we borrow terms from our own social structures. Despite all the profound differences, social systems of bees and humans often follow similar rules for problem-solving, yet those in bees stem from natural selection, whereas those of human societies primarily originate from cultural evolution. Although some mechanisms may be similar, others will be completely different, and any comparisons are only helpful if they facilitate comprehension in either system.

The scientific progress made by students of honey bee biology in the future will show where we have erred today. Honey bee research, as with any research, is driven not just by finding new principles but also by unraveling errors in already identified processes. Truth becomes conditional: It may be true for some time but eventually be proven false. Never trust a scientist who tells you that he or she knows the truth. Almost all scientists in history have eventually been proven false on some account as time went by. The scientist proven to be wrong may often be more important to a developing understanding of natural phenomena than contributions that have provided what is generally accepted to be the truth today. The asymmetry in relation to scientific hypotheses—that they may be falsified but not proved true—generates difficulties for those who view natural science from outside its domain. The expectation in today's societies that the natural sciences will provide unequivocal answers does not take the provisional nature of scientific evidence into account, often with unfortunate results for the public and its political leaders. Alas, policymakers cannot delay making decisions based on accepted knowledge that might subsequently be rejected. Instead, they move forward making "evidence-based" decisions, and history eventually shows whether these were correct or false. However, the veracity of evidence needs to be carefully evaluated, such as the "fact" of the sun circling the earth. Evidence-based scientific reasoning is highly susceptible to ethical, religious, economic, and political constraints. History has demonstrated that science does not operate in a policy-free environment. Science can only be as good as society allows it to be. In addition to being misguided by "common sense" or misfunded by a seemingly policy-relevant but scientifically irrelevant call for a tender, the scientist may be trapped by his or her own results, obtained in good faith and carefully following state-of-the-art procedures. This is the consequence of today's scientific philosophy of accepting a result if it satisfies the rules of statistical significance. If the level at which a hypothesis is rejected, although it is actually correct, is less than 5%, the scientist is satisfied and concludes that the result is valid and is based on a sufficiently large sample. Yet, do not be fooled, and keep the following in mind: With an error rate of 5%, 5 studies in 100 identical empirical studies make exactly this error—not because this was faulty science but simply because an unlucky sample yielded a significant result. The problem is that we do not know which are the 5% false-positive results out of the 95% correct ones. Very likely, despite having been alert when searching for support of our arguments in this volume, we cannot exclude having been trapped in these inevitable pitfalls of science ourselves.

Nevertheless, our exploration of the complexity and idiosyncrasies of honey bee social organization is intended to provide you with a fresh set of insights and novel questions that will have the effect of revealing the presence of false kings and provide a greater depth of understanding of evolutionary possibilities in these highly social organisms. We hope that these scientific excursions raise your critical awareness when trying to get a better understanding of the natural world of the honey bee.

{ 2 }

Out of the Dark

2.1. From wasps to bees

Yes, there was a time before the bee, but you would not have liked it. It was a dark world without flowering plants competing to attract buzzing bees for pollination—a world without colorful fruits and berries. Even after solitary bees emerged, it took many millennia before bees became social and a colorful and sweet world, a place with nectar, emerged. Today, many of us believe that the world has been molded in most radical ways by recent human activities (Harari 2014). Yet our planet experienced far more fundamental changes in the Early Cretaceous era, 145–113 million years ago (Mya), when flowering plants and their associated bees emerged for the first time (Figure 2.1). Indeed, plant–pollinator interactions were the central evolutionary innovation that changed the functioning of ecosystems and set the stage for a tremendous radiation in plant and animal diversity. This step was also essential to the success of vertebrates, eventually resulting in the evolution of humans.

This collaboration between evolving eudicots or flowering plants and early bee species is thought to have started in the relatively dry regions of Western Gondwana approximately 120 Mya (Cardinal and Danforth 2013). Over time, it created a beautiful variety of flowering plants and their associated pollinators, the majority of which are bees. Understanding the evolutionary origin of honey bees is therefore key to a range of insights into their current state of existence and allows us to make educated predictions about their future survival, particularly in a world shaped by the impact of an increasing human population. The honey bees, which are highly social, are one small branch of an evolutionary tree that has produced an astonishing variety of bee species with a wide range of social complexity from solitary species through various forms of social existence to the highly eusocial stingless bees, bumblebees, and honey bees (Cardinal and Danforth 2011).

So, where did the bees come from in the first place? Phylogenies of wasps and bees suggest that some ancestral wasp species gradually transformed into

FIGURE 2.1 *A worker of the giant honey bee Apis dorsata meeting a worker of the dwarf honey bee Apis florea.*

a proto bee by switching from a carnivorous diet to a diet exclusively of plant origin made up of nectar and pollen but retaining the behavior of provisioning their helpless brood with food they collected. This change in diet required important morphological changes, resulting in the development of specific pollen collecting devices that are common to all bees (see Chapter 3). Since then, bee diversity has blossomed in much the same way as has their floral hosts, with as many as 29,500 bee species currently in existence and an untold number of extinct species from which they arose during the past 120 Mya.

Understanding the marvelous adaptations of honey bees should be accompanied by the appraisal of the dark side of the honey bee colony with its jerry-built evolutionary solutions. We explore the phylogeny of the bees to identify the relevant timescales and interpret the species radiations through which the primitive bees of Gondwana eventually transformed into the honey bee and their complex societies. We not only travel along the phylogenetic road that may have been followed by the honey bees but also explore whether it is more likely to be a road out of the dark times into a brilliant future or rather one into the dark dead end of extinction.

The quest for the origins of honey bees has been bedeviled by fragmentary fossil evidence; hence, there is no clear understanding of the historical distribution of ancient species of honey bees that could be the ancestors of modern species. This is a beautiful setting for a controversial academic debate. The ancestral switch from carnivorous wasp to a strictly vegan bee can be reasonably timed to the arrival and diversification of the flowering plants. However, when they went on to evolve distinct female castes with divergent morphologies remains a conundrum to be illuminated.

The sparse fossil record of the bees is based on specimens found in amber of various ages (Poinar and Danforth 2006). Only 244 fossil bee species have been described so far, of which 24 species have been identified as ancestors of the bumblebees and honey bees, and 17 of them are described as being in the genus *Apis.* With such a sparse fossil record, it has been difficult to extract a reliable trace of the evolutionary pathway from the wasp-like ancestral solitary bee that eventually ended up giving rise to eusocial bees that are characterized by their complex colony-based life history and highly sophisticated social behavior (Michener 1974). Because most of the fossils of the social bees are workers, it seems most plausible to infer that eusociality must have evolved earlier than the age of the particular fossils. The oldest fossil social bee that has been discovered—a worker stingless bee species, *Cretotrigona prisca* (Engel 2000)—dates back to 60 Mya (Figure 2.2). Sociality in bees is at least that old.

Today, most bees pursue their existence in solitary fashion, with females raising helpless brood on their own, young males and females emerging from these nests, mating, and the females then starting a new cycle of brood production for the next generation. Some bee species have evolved various forms of communal or collaborative living that has given rise to more complex societies (Michener 2007). However, the iconic bees that are favored in human stories and legends are not these obscure solitary and weakly social species but, rather, those for which there has been a dramatic change in social organization to produce complex societies with large numbers of individuals and high levels of cooperation. This group of exemplary species (380) living in well-structured colonies represents only approximately 1.3% of all bee species. Although they are individually abundant and their complex behavior is regarded as a marvel of evolution, they are rare in terms of species numbers. This is most extreme for the genus *Apis*, for which only 11 different honey bee species have been recognized to date (Table 2.1), with most of the species diversity (10/11) in eastern Asia. Some taxonomists divide the genus into three subgenera: the giant honey bees *Megapis*, the dwarf honey bees *Micrapis*, and the medium-sized cavity-nesting honey bees *Apis s. str.* However, this does not contribute to explaining the small number of honey bee species.

So can we interpret today's low species diversity in an evolutionary context? Is the genus *Apis* only recently differentiated and we are witnessing just the

FIGURE 2.2 *A reconstruction of the fossil stingless bee species, Cretotrigona prisca.*

TABLE 2.1 Extant Species of Honey Bees and Their Distribution

Species	Location	Nest Type	Worker Size
Apis florea	Arabian peninsula to Southeast Asia and China	Exposed comb	Dwarf bees
Apis adreniformis	Southeast Asia and China	Exposed comb	Dwarf bees
Apis dorsata	India to Southeast Asia	Exposed comb	Giant bees
Apis laboriosa	India to Laos—high altitude	Exposed comb	Giant bees
Apis breviligula	Philippines	Exposed comb	Giant bees
Apis binghami	Sulawesi	Exposed comb	Giant bees
Apis cerana	Widespread in Asia	Cavity nesting	Intermediate size
Apis koschevnikovi	Malay Peninsula, Borneo, Brunei, Java, Sabah, Sarawak, and Sumatra	Cavity nesting	Intermediate size
Apis nigrocincta	Sulawesi	Cavity nesting	Intermediate size
Apis nuluensis	Sabah	Cavity nesting	Intermediate size
Apis mellifera	Europe, Africa, and the Middle East	Cavity nesting	Intermediate size

beginning of a tremendous radiation within *Apis*? Or is the opposite true—that they are an old, established genus whose species diversity has declined during the past 30 million years? The fossil evidence suggests that the genus has been more diverse in the past and that individual species have always been widespread in Europe, Asia, and North America (Kotthoff, Wappler, and Engel 2013). The impressive abundance of the orchid bees and the vast abundance of highly efficient stingless bees in South and Central America may well be associated with the absence of endemic *Apis*-type bees in those regions.

Whereas honey bee species make up 7% (17/244) of the bee species represented in the fossil record, they represent only 0.0003% (11/29,500) of the extant species of bees. One might argue that the likelihood of fossilization varies substantially with a bee's lifestyle and that social species with large colonies are more likely to be fossilized. However, even within contemporary social bees, the number of honey bee species is extremely low compared to the numbers of stingless bees and bumblebees. Even more surprising, the number of fossil species is actually highest for the honey bees, suggesting that many more honey bee species than other social bee species may have existed approximately 30 Mya. Thus, although honey bees are common from a perspective of global abundance today, they are actually extremely rare from the perspective of species biodiversity. They represent the smallest genus of all genera in the large clade of bees, and they represent only 3% of all eusocial bee species. What is particularly puzzling is that their species diversity is much lower than that of the other groups of eusocial bees with large colonies, the stingless bees and the bumblebees. If the fossil record gives us a correct insight, this may have been very different in the past. During the Oligocene, the genus *Apis* was much more diverse than any other genus of social bees.

The relative lack of species diversity within the genus *Apis* may be related to the fact that the species, especially the cavity-nesting species, have very large geographic distributions and mating systems that do not lead to isolated populations in which speciation could occur (see Chapter 5). Today, the biodiversity of *Apis* species seems to be reflected in their subspecies. The western honey bee, *Apis mellifera*, has been classified into a bewildering array of subspecies, often representing local population differentiation within an extremely diverse species. The same phenomenon can be observed for the Asian cavity-nesting species *Apis cerana*. Across its huge distribution range spanning from cold temperate to tropical climates, a huge variety of subspecies have been identified (Hepburn and Radloff 2011). Some of the fossil evidence suggests that extinct species were also widespread and were highly variable, as is found in *A. mellifera* today, making it difficult to classify them as true species or subspecies of a highly variable fossil species (Figure 2.3).

Nesting behavior in the the honey bees follows two distinct patterns: There are the open-nesting species with exposed combs seen in the giant and dwarf

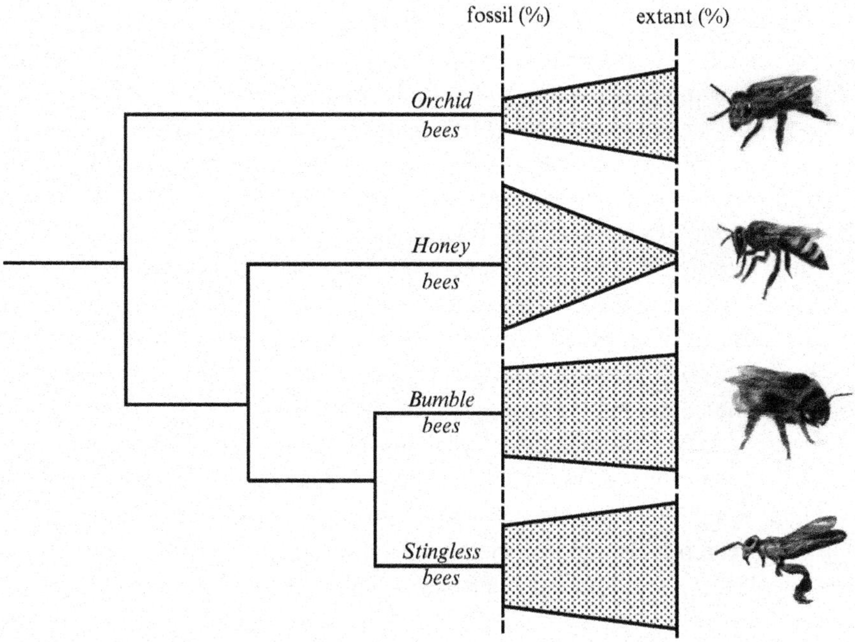

FIGURE 2.3 *Phylogeny of the eusocial bees and their solitary ancestor (drawings not to scale) based on the analysis of Romiguier et al. (2016). The length of the sides of the trapezoids reflects the relative frequency (in percent) of bee species known in the various bee genera from the fossil record (left, 40 species) and the extant number of species (right, >761 species).*

bees (Table 2.1) and the cavity-nesting species in the bees of intermediate size. Exposed combs place constraints on the environments in which the colonies can survive, with the species being confined to tropical and subtropical climates. The cavity-nesting species are able to place multiple combs within the cavities that they occupy (see Chapter 8) in hollow trees, cliff faces, or the ground and to insulate themselves from extreme weather conditions. The cavity-nesting species have also developed the ability to control the temperature within the cavities so that they can maintain a relatively uniform temperature for brood production. These adaptations allow them to survive outside of the tropics where winter conditions are much harsher. In addition, the provision of cavities for the bees to nest in was the start of beekeeping activities by humans.

2.2. Out of the Northern Hemisphere

The fossil record does not give us many clues as to the way in which the honey bees acquired the particular set of traits that characterize their form of eusocial organization. However, the fossils, which are all identifiable as eusocial honey bees because they are workers (the most abundant caste), can be used to provide a scenario for the route, both evolutionary and geographic, by which the genus *Apis* may have arrived at the geographic distribution of current honey bee species. The problem, however, is again the extraordinarily small fossil sample size, which allows for many alternative interpretations. The oldest honey bee fossils (25 Mya) represent a set of highly variable individuals that can be grouped into two major morphotypes: one that is similar to the giant honey bees and and one similar to the cavity-nesting types. Both types have been found in Europe. This is good news and bad news at the same time. Because most fossils have been found in Europe, it is slightly uncertain whether this is due to sampling intensity and the availability of fossil sites harboring honey bees or whether they were more abundant in Europe than elsewhere.

The clear point is that honey bees of *Megapis* and *Apis* have been detected on the same continent, with only the cavity-nesting *A. mellifera* surviving to the present. So it is clear that the subgenus *Megapis* became extinct in Europe, most likely during the Quaternary ice ages. Indeed, cold winters may be detrimental to open-nesting bee species. *Megapis* species were present 25 Mya in the Oligocene but also approximately 15 million years later in the Miocene in Europe. *Megapis* fossils found in Japan also date back to the Miocene, suggesting it was a common group of species on the Eurasian continent. The discovery of the fossil *Apis nearctica* honey bee in Nevada suggests that *Apis* honey bees in the Miocene were abundant throughout the Northern Hemisphere, including Eurasia and America (Figure 2.4).

Rather than trying to infer or comment on the many sometimes controversial scenarios on honey bee biogeography that one can draw from the rather

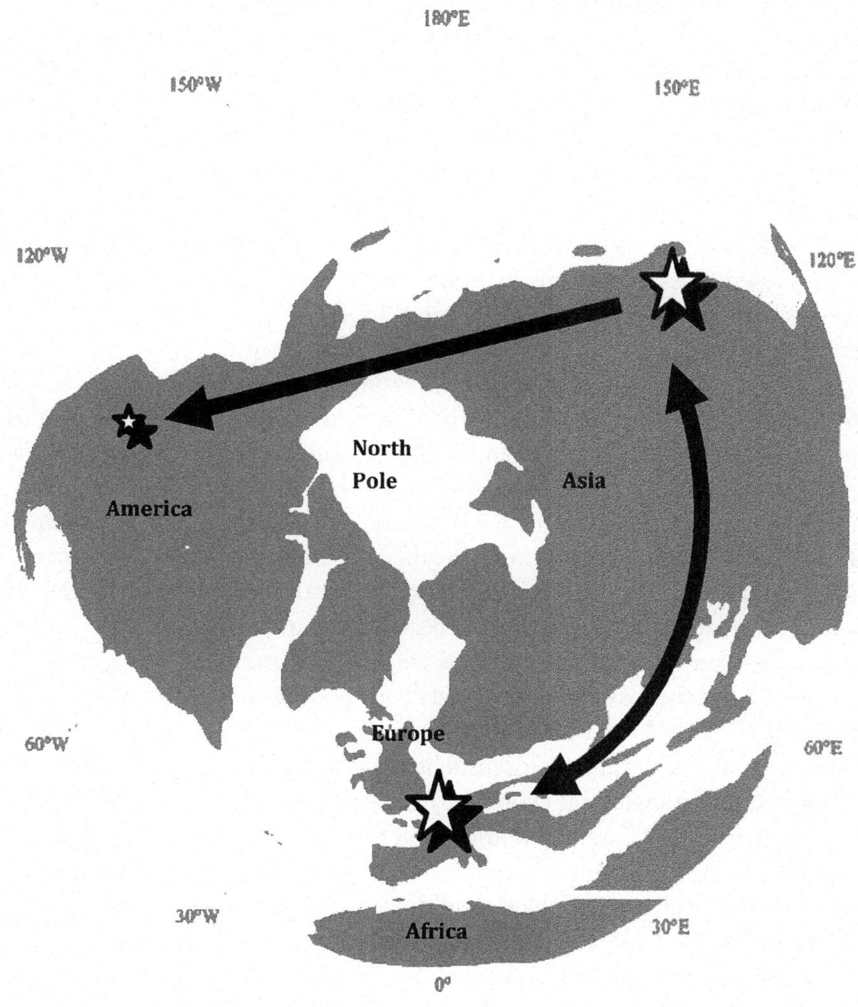

FIGURE 2.4 *The distribution of honey bee species inferred from the fossil record during the Oligocene in the Northern Hemisphere. The stars indicate the locations of fossil Apis specimen finds in amber or shale.*

sparse fossil record, we draw attention to the two obvious large-scale extinction events. The first one is the extinction of all *Megapis* species from Europe, and the other one is the extinction of *Apis* from America after the Miocene. Climatic conditions may have had very drastic effects on any open-nesting honey bees, such as the *Apis dorsata*-like *Megapis* species. The Miocene was warmer than today, with a warm temperate climate throughout most of Europe. It is known from the fossil record that there were palm trees and alligators in northern and western Europe (even in England), and an open-nesting honey bee species might well have enjoyed this kind of climate. In the Oligocene, it was just as warm, with a paratropical climate in western and central Europe (Figure 2.5). The climate was equally honey bee friendly in northern America. Eurasia was connected to America by Beringia, a massive land bridge between Alaska and the Chukotka peninsula spanning today's Bering Strait. During both the Oligocene and the Miocene, this region had a cool temperate climate, very likely allowing for the spread of honey bees from Asia into America, certainly if they were cavity nesting.

The subsequent cooling of the planet and the recurrent ice ages may well have caused mass extinctions of honey bees. In particular, the open-nesting species would have been affected most. Today's open-nesting species are constrained to tropical and subtropical climates. In contrast, the cavity-nesting species can sustain long periods below freezing, so *A. mellifera*-like bees should have been able to survive south of the European mountain ranges during the periods of peak glaciation, as did the many bumblebees and many other solitary bee species.

Given that we are aware of the extinctions of the entire *Megapis* clade in Europe and the potentially cavity-breeding *A. nearctica* lineage in America, the radiation of the genus has been highly constrained. Only 11 species are currently recognized, which seems to be a particularly low number by comparison with other social bee species, and only two of them have large geographic ranges. The sparse fossil record suggests that species abundance may have been higher in the past. If we were pessimists, we would argue that the loss of species diversity in the genus and their sensitivity to climatic changes puts them on the reddest of all International Union for Conservation of Nature Red Data Lists. If we were optimists, we might argue that this is just the beginning of a great honey bee age. We are at the very beginning of a radiation taking place within the western honey bee. So which scenario is more likely?

Although the fossil honey bee data do not date back as far as those of other bees, it seems the pessimists may be right. The fossil record of the social stingless bees, with 11 species, is similar to that of the honey bees, yet there are currently well over 300 stingless bee species distributed across the tropics of the New and the Old World. The ratio between recognized extant and fossil species is more than an order of magnitude higher in the stingless bees (see Figure 2.3). Their radiation has been enormous. This is similar for fossil (11 species) and extant bumble bees (>250 species). The bias is even more extreme for the closely related orchid bees, which are only represented by three fossil species compared to an impressive 200 species endemic to the neotropics.

So the genus *Apis* seems to be in a parlous state from a biodiversity perspective (see Figure 2.3). With a fossil record of 17 species versus 11 extant ones, a significant amount of diversity must have been lost. If we take the ratio between fossil and extant orchid bees, *Apis* should be represented by approximately 800 honey bee species today. The price that *Apis* species appear to have paid for their ecological success in terms of abundance and extensive distribution has

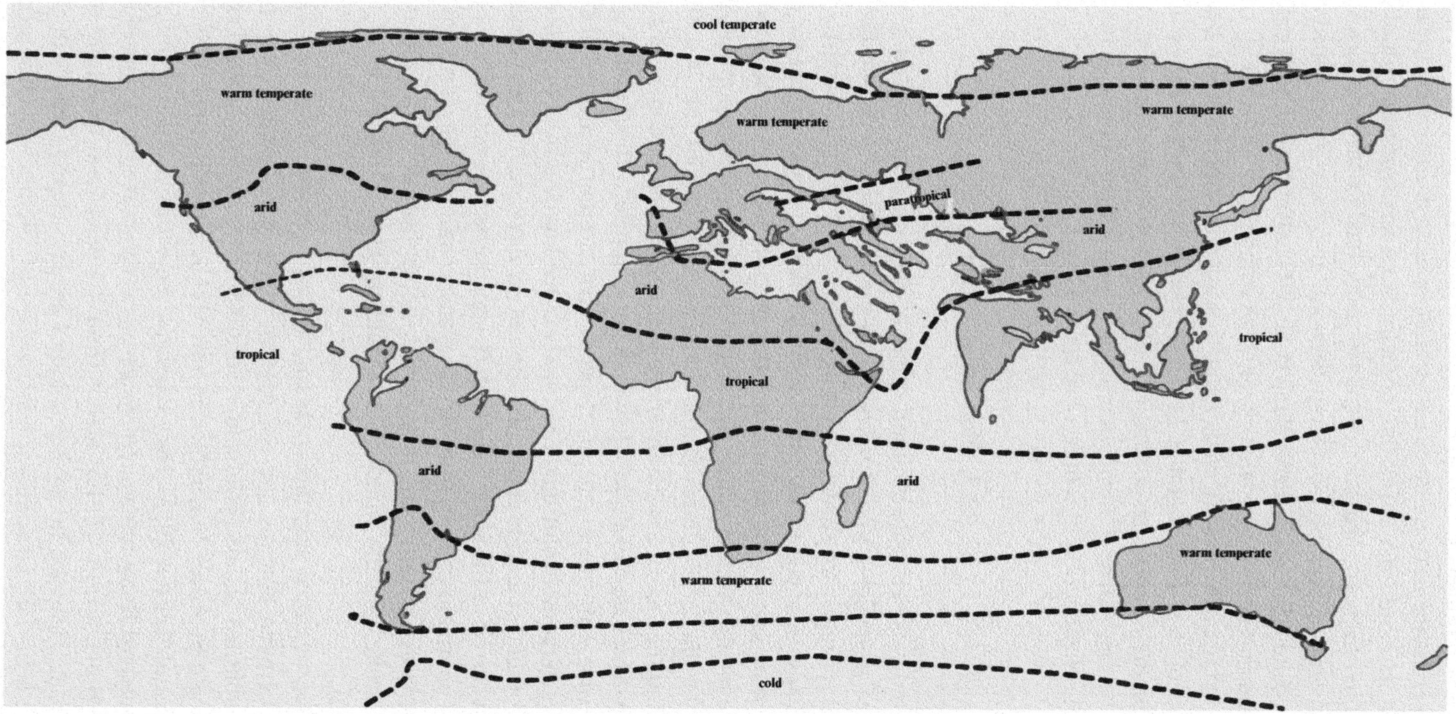

FIGURE 2.5 *Climatic zones on the continents during the Oligocene indicating the benign climates at northern latitudes when honey bee species were becoming widespread.*

been a dramatic impairment of biodiversity at the species level. Given the enormous success of the western honey bee *A. mellifera* and its increasing use in apiculture (Moritz and Erler 2016), the future for the Asian species may look even bleaker. *Apis cerana* populations in China have declined by 60% and their distribution has been reduced by 75% (Yang 2005) and they have also declined in the Japanese islands (Sakagami 1959; Yang et al. 2011) after having been replaced by *A. mellifera* (Oldroyd and Nanork 2009). A small ray of hope for *A. cerana* can be found in the expansion of its range to Australia (Koetz 2013b). Because the number of *A. mellifera* colonies kept by beekeepers throughout the endemic distribution of *A. cerana* has dramatically increased in the past decades, it may very well be that global *A. mellifera* apiculture rather than climate change will cause the next extinction of a honey bee species (see Chapter 9).

2.3. Not out of Africa

The enormous endemic distribution range of the two cavity-breeding honey bee species *A. mellifera* and *A. cerana* with a huge diversity of subspecies is rather extraordinary for eusocial bee species. In fact, it partially resembles that of *Homo sapiens*, a species that exhibits similar characteristics in that the genus has been more speciose in the past, but this species diversity has been reduced to a single highly variable, widely distributed species with large populations. It may well be the fierce competition among generalist species that is a driving force behind the lack of species diversity. Honey bees and humans can survive almost anywhere on the globe. They may not compete with specialist species but may well with other generalist species that overlap in their niche. The similarities in recent evolutionary history between humans and honey bees are indeed profound. The subspecies radiation of *A. mellifera* has been estimated to have occurred at the same time that *H. sapiens* became established on the European continent coming from Africa (Stewart and Chinney 2015) 20,000–51,000 years ago. During this time period, honey bee subspecies radiated in Africa, Europe, and the Middle East (Wallberg et al. 2014; Fu et al. 2016; Figure 2.6).

The biogeographic migration patterns, however, seem to have been slightly different between humans and the western honey bee. Despite some prominently published claims (Whitfield et al. 2006), *A. mellifera* most likely did not originate in Africa but, rather, in the Middle East (Ruttner 1988; Han, Wallberg, and Webster 2012). How the displacement of competing honey bee species occurred is not known. *Homo sapiens* is believed to have simply killed its competitors, a trait still not absent from the behavioral repertoire of the species today. However, honey bee colonies are also clearly capable of rather aggressive behavior and readily rob and kill other colonies when resources are scarce and there is a need to save their own colony (see Chapter 4). Whatever

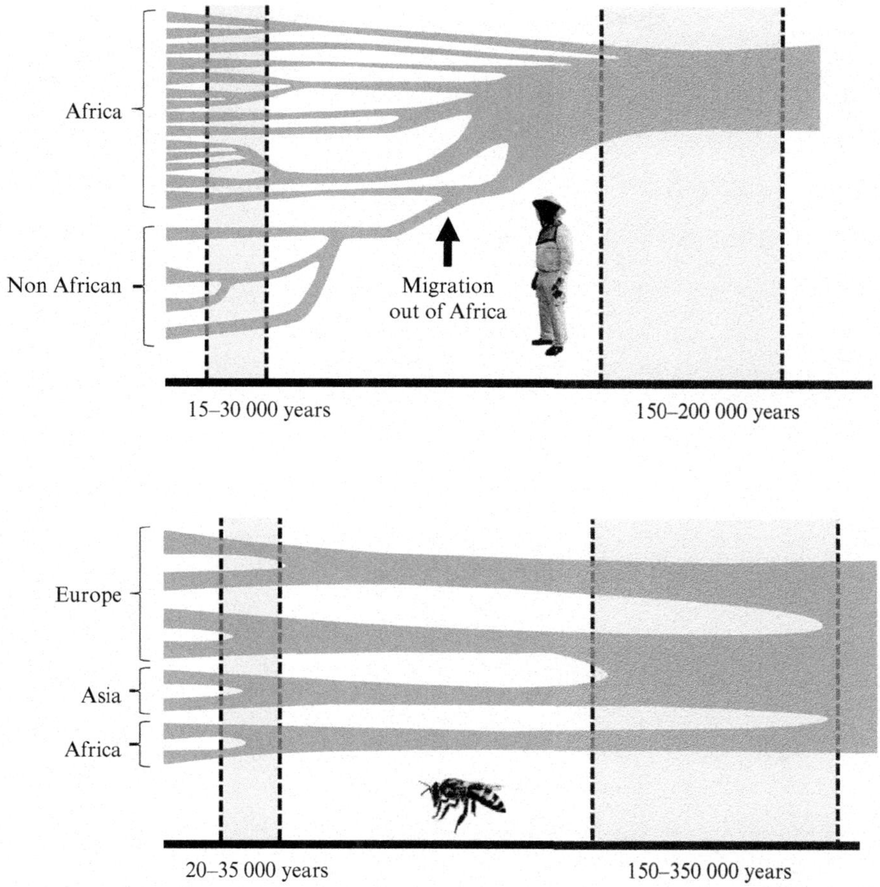

FIGURE 2.6 *Concurrent biogeographic radiation of the honey bee populations inferred from genome sequence variation over the recent past (bottom) compared with human population radiation (top) over roughly the same time period of 20,000–35,000 years ago (Wallberg et al. 2014; Fu et al. 2016).*

the driving forces may have been, the result is rather clear: Only a single species with a broad genetic and phenotypic variance survived. Today, it seems the evolutionary fates of these two species are now inextricably intertwined. Humans cannot survive without honey bees, and honey bees in the industrial world depend for their survival on humans.

{ 3 }

A Difficult Diet

3.1. The vegan honey bee

Honey, by far the most popular bee product, is not only a cherished food but also used in a variety of treatments for diverse ailments. Consumers expect their honey to have high quality and purity. Contamination of any kind may be particularly detrimental due to its wide use in "home medication." Like any other agricultural product, we trust quality control in the food chain from the site of production to the end user. Indeed, the control of food processing "from farm to fork" is a major tenet of modern agricultural and consumer health policies. In this case, we generally trust the beekeeper, but we certainly fully trust the honey bees for their competence in food processing and the production of pure honey. So is our trust in the competence of the honey bees warranted? It may help to understand why they produce this honey in the first place. It is certainly not to make humans happy so they can harvest the bees' honey. A broad range of antimicrobial compounds are found in honey, whose primary function is to contribute to the health of honey bees rather than humans (Erler and Moritz 2015).

Honey bees face a dietary problem because they are strict vegans and live exclusively on plant products that provide the carbohydrates and proteins they need. This is an evolutionary ancient trait as almost all bee species from solitary to social share this diet. Bees have to deal with exceptionally difficult types of forage: They need to handle and transport liquids (nectar and water), dust (pollen), and glue (plant resins). To deal with these materials, they have a tank for liquids (honey stomach; Figure 3.1) and a dust broom to collect and press pollen grains into pellets that can then be transported in the pollen baskets on the hind legs (corbicula) (Snodgrass 1956). They also use the pollen baskets to transport the sticky plant resins used to make "propolis," a building material used to seal cavities inside the nest.

These peculiar dietary and morphological adaptations for collecting their food are the major adaptive features of honey bees. Efficient use of these

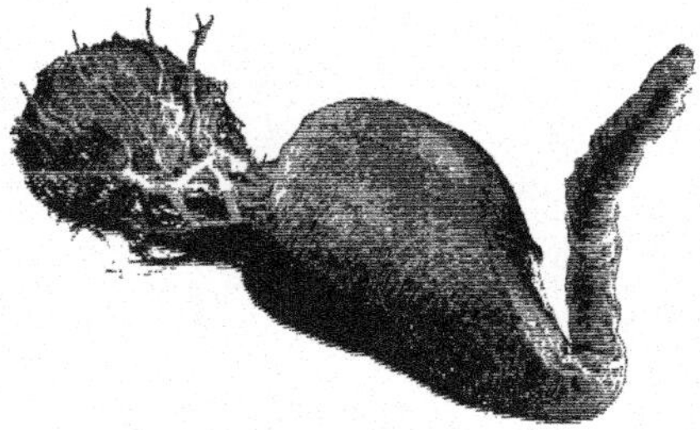

FIGURE 3.1 *The tank (honey stomach) a honey bee worker uses to transport liquids (nectar and water).*

tools is dependent on the use of their communication skills and the collaboration of workers in the hive. In order to fill a jar with 1 kg of honey, a single worker with an average load of 25.3 mg (Lundie 1925) would have to conduct close to 400,000 foraging flights covering a distance equivalent to a return flight to the moon and back to harvest the required quantity of nectar. Fortunately for the poor worker bee, she is not alone in this task. Of course, it is the concerted effort of the foragers and in-hive workers in the colony that handle the food intake. Foragers collect the nectar, deliver it after arrival to receiving workers, who in turn store it in the honeycombs. However, bringing it to the colony is one issue, storing it is another one altogether. The honey bees need to store their food over very long periods of time to bridge periods of dearth, including long winters, without having a deep freeze. It requires major logistics and food processing to solve this problem. Processing is less critical for the stored pollen grains, which protect themselves against fungal and antimicrobial infections by the exine, a sturdy pollen wall (Kerstiens 1996). This is not the case for nectar, which is prone to rapid fermentation if not processed quickly and properly. Nectar is primarily composed of water with a variety of sugars at different but always low concentrations depending on the plant species. The low sugar concentration is important because nectar has to be highly liquid so that the bees can easily suck it from the flowers. However, low-sugar solutions have a severe drawback: They cannot be stored for long because they are particularly prone to fermentation by microbes unless kept refrigerated. Alas, evolution had not foreseen the development of refrigerators in the hive. On the contrary, the conditions in the colony provide an almost perfect setting for microbial fermentation with an environment at 30°C and 60% relative humidity. The low sugar concentration of nectars provides an excellent substrate for both aerobic and anaerobic bacteria.

Although some plant nectars contain traces of antimicrobial substances and the honey bee workers add antimicrobial compounds to the nectar during processing, these are at too low a concentration while the nectar is dilute. In-hive workers therefore transform the liquid nectar into a sticky syrup (honey) by actively evaporating the water. They do this by exposing small nectar droplets on the tongue in an airstream generated by wing fanning. High colony temperature and forced air ventilation by fanning in the colony further contribute to the lowering of the water content of honey in the storage cells. Any incoming sucrose in the nectar is digested by the workers into fructose and glucose, which are the major sugars in honey. This digestion is critical to reducing the likelihood of crystallization of the glucose in the honey (Doner 1977) because the fructose produced by splitting sucrose (into glucose and fructose) helps keep the glucose in solution as the nectar is concentrated (Wright, Nicolson, and Shafir 2018).

In the end, the total sugar concentration in the final product is raised to approximately 80% (Crane 1975). From a food quality perspective, this is convenient for long-term storage because the high sugar concentrations prevent fermentation of the food for many months. Honey bees do well storing their food without a refrigerator.

3.2. Sticky honey

Whereas high sugar concentrations are fine for storage, they have a serious disadvantage when it comes to handling: The sugars make the honey highly viscous and very sticky, which can be absolutely lethal to the bees if they contaminate themselves with it. A worker covered in honey dies quickly because the spiracles are blocked and it cannot continue tracheal breathing. So honey bees ensure through strenuous grooming that they avoid contact with sticky honey, and the incentive to keep the hive clean is exceptionally high for every worker. Any spot in the colony or any bee contaminated with honey is quickly groomed by nestmates. They dilute the honey and take it up into their honey stomach to consume it themselves, feed it to others, or deposit it into the honey stores. Cells filled with mature honey are sealed with a wax cap that, among other things, also prevents uncontrolled contamination of the bees in the colony.

3.3. Dusty pollen

Whereas honey contamination poses a food handling problem, pollen as a protein source is also problematic. It is difficult to handle; just consider your own problems in handling dust without a vacuum cleaner. Honey bees are able to manipulate pollen only because they are exceptionally clean insects and try to get rid of any contamination on their body surfaces. The pollen combs on their hind legs are their dust brooms, and these work very effectively to gather pollen contaminating their bodies. The sticky pollen grains attached to the hair of the body surface are removed with these combs, and the pollen can then be compressed into pollen pellets that are stored in the corbiculae of the hind legs (Figure 3.2). So far so good. Pollen is the only protein source for bees, and although evolution has shaped a most efficient collecting apparatus for its acquisition, its digestion presents a serious challenge. The thick indigestible exine of the pollen grain needs to be eliminated before the nutrients can be accessed.

So what to do with the exine once the nutrients are extracted? In the adult bee in summer, this is not a problem because the pollen walls can simply be excreted though the alimentary canal. The workers fly out to defecate. However, what to do in winter or those "British" summers with extended periods of rain? This is much more critical because the bees cannot fly out and their hindguts will fill up with pollen grain husks. Defecation inside the colony is not an option for healthy bees. Remember, honey bees are clean insects! The workers go to extraordinary lengths to keep the colony clean to prevent the spread of gut parasites and pathogens. Yet long winter periods are more relaxed than one might think. Protein is primarily needed for rearing the brood. Hence, it is the nurse bees that have a high protein requirement and the queen who is producing the eggs. In winter, this is different. Brood production stops, and the need for

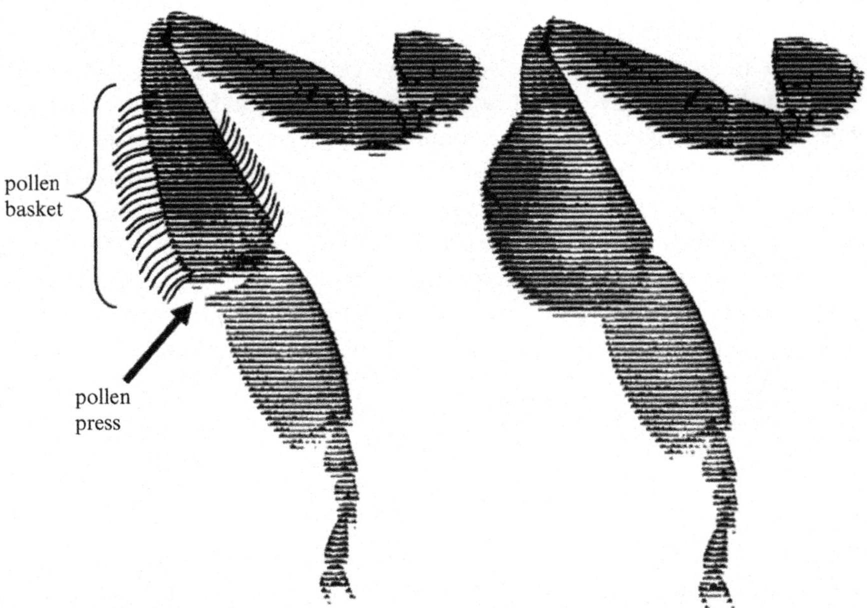

FIGURE 3.2 *Pollen collection and transporting mechanism. Pollen grains are brushed by comb-like hairs into the pollen basket, and press-like leg structures compress the individual pollen grains into a solid pack of pollen firmly attached to the leg (right).*

anabolic metabolism is almost completely absent. The adult winter worker does not have any need for protein metabolism. There is not much brood present (if any), so the food glands do not need to be developed and the colony only needs to ensure that the nest temperature is maintained such that all bees survive. Regulating temperature only requires energy; thus, honey is consumed. Protein metabolism is drastically reduced in old adult bees. They do not need to develop new tissue—it is all about energy metabolism either during flight or during colony heating. The respiratory quotient (ratio between carbon dioxide output and oxygen consumption) is therefore close to 1. Nevertheless, honey is not completely free of pollen grains, and eventually the bees need to fly out in order to defecate. This is a particularly dark side of the hive for automobile owners, who should not park their vehicles in front of an apiary on the days of cleansing flights.

For larvae, the issue is quite different. For them, pollen is a truly troublesome diet. They have a closed rectum and cannot defecate until pupation, which seems adaptive because it prevents defecation into their own food (Figure 3.3). Hence, any pollen exines they ingest must stay in the gut for more than 6 days until they complete larval development. Only after the midgut and the hindgut have fused just before pupation will the larvae defecate. It therefore seems reasonably adaptive that the larvae are fed primarily with processed food in the form of a secretion of the hypopharyngeal glands that is rich in proteins and

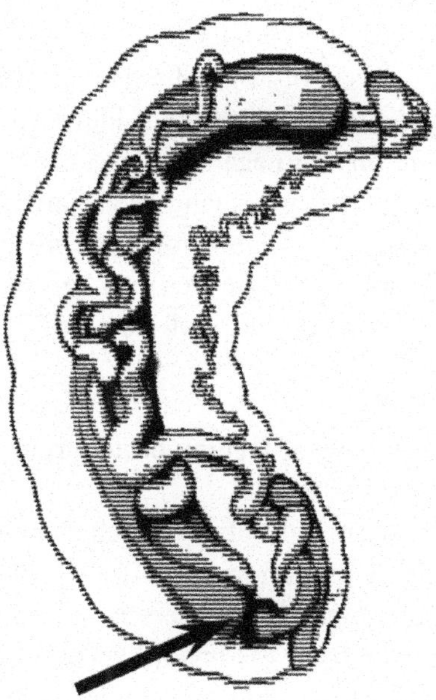

FIGURE 3.3 *The larval rectum is not connected to the hindgut (arrow). Defecation is only possible during pupation; hence, indigestible food items are a burden during larval development.*

fatty acids and minimizes the need for proteins derived directly from pollen gains and hence the generation of a pollen load in the midgut.

Pollen, however, has another downside. If kept at 35°C, it quickly loses its nutritional value for the bees. Honey bees therefore process the pollen into the so-called bee bread. The bees cover the pollen with honey and glandular secretions that conserve the nutritional value and reduce the germination capacity of the stored pollen (Herbert 1992). However, the honey bees have help handling the problem. Pollen is full of a variety of fungal spores (Gilliam, Prest, and Lorenz 1989), and most of these are highly beneficial. They contribute to the efforts of the honey bees by adding antibiotic compounds and many other compounds that enhance the nutritional value of the stored pollen (Yoder et al. 2013).

Thus, the food brought in by the foragers goes through many "hands" (mandibles) before it is ready to be fed to the members of the hive. Given that during every processing step the workers test the quality of the honey and pollen, it seems quality control is rather intensive. Food conservation efforts are also efficient, and the honey bees seem to have an excellent "farm to fork" quality control chain. Of course, the most drastic quality control is when the nurse bees start producing food jelly in their food glands. Here, the honey and pollen are metabolized to a sterile high-quality food that is fed not just to the larvae but to all members of the hive (Crailsheim 1991, 1992). The nurse bees are in a particularly central position for food processing in the hive. They have the highest potential for protein digestion among all bees in the hive. They serve as the digestive caste in the colony, processing the pollen proteins to food jelly that can be fed to all, brood and adults. So food quality control in the hive is exceedingly stringent with a series of redundant inspections. The highly proteinaceous food jelly produced by workers is indeed a marvel of evolutionary adaptation to overcome all the negative aspects of the original nectar and pollen supplied by plants. Despite having tried hard, we have not found any maladaptations here: Division of labor for food processing works excellently and with high efficiency among the various members of the colony, as has been meticulously demonstrated by the work of Karl Crailsheim and collaborators.

3.4. Demands of feeding brood progressively

Feeding larva is the most critical nutritional challenge in the colony because they are highly sensitive to deviations from the required diet. Their primary source of food is a protein-rich liquid secretion termed food jelly (or royal jelly if fed to queen larvae) from the hypopharyngeal and mandibular glands of the nurse bees. However, in addition to this high-quality secreted diet, a less quality-controlled mixture of honey and pollen is added to the larval diet of workers and drones, depending on the developmental stage of the larva. The feeding of the larvae is done in a progressive manner with the diet being adapted to its developmental stage. This mode of brood care has been termed "progressive feeding." The specific adaptations of workers for efficient progressive feeding are extraordinary. The nurse bees serve as the liver of the hive by transforming the pollen protein into food jelly. Not only do they feed the larvae at the right time with sufficient food but also they very carefully regulate the quality of diet the larvae receive. After hatching from the egg, the larvae are fed on a diet consisting exclusively of secretions from the hypopharyngeal glands of the nurse bees. The composition of this larval food has been studied in exceptional detail, particularly because it is also the diet used to raise honey bee queens, which are fed exclusively on this secretion (royal jelly) throughout their larval development (see Chapter 5) (Buttstedt, Moritz, and Erler 2014). In the case of larvae destined to be workers, their diet is altered so that they receive a mix of food jelly with pollen and honey, which delays their development and reduces their growth to ensure that they become workers.

In contrast to the honey bees, allodapine bees, and a few bumblebee species, most highly eusocial bee species are so-called "mass provisioners." They deposit a mass of mixed honey and pollen into a cell before the queen lays an egg on top. The cell is then sealed, and larval and pupal development occurs within the cell without further interference from the workers. Honey bee workers, on the other hand, continuously interact with larvae during the period of progressive feeding of the brood and seal the cells only shortly before pupation. This behavior is rather rare for the vegetarian bees and is more typical of eusocial predatory wasps and ants that cannot store their food. Their larvae feed on animal protein from prey, which needs to be fresh because it cannot be stored in the nest. It would quickly decay as a consequence of bacterial and fungal contamination and be a source of potential infections. Bees do not have this constraint because they process their food to preserve its quality even under the high-temperature and high-humidity conditions in the hive.

So why do honey bees make the effort to progressively feed their offspring when they have these precious food stores? Why not choose the ancestral type of mass provisioning? It seems so much "easier" to simply fill a cell with the correct mix of honey and pollen, let the queen lay an egg, seal the cell, and forget about it. Actually, it would even reduce the risk of spreading brood diseases within the colony because there is absolutely no contact with the developing larvae (see Chapter 7). Field (2005) modeled some conditions under which progressive provisioning could evolve in nonsocial insects. These included the risk of mortality from non-inspected larvae due to parasitism. However, this risk is negated in honey bees because they can detect diseased or parasitized pupae even in the sealed cells. Workers are known to express "hygienic behavior" in which they selectively open sealed brood cells if they contain infected pupae. These pupae are then removed from the cells and discarded from the colony.

The value of progressive feeding might be enhanced, however, if the brood developed in a way that facilitated increased fitness at the colony level. Despite the higher risks of spreading diseases in the colony, there might be benefits if more brood could be raised in a shorter period of time than could be achieved by mass provisioning. Certainly, the growth of the colonies after winter or other periods of dearth is essential for the honey bees to exploit the sudden mass nectar flows in spring or after desert rainfalls. Indeed, the developmental time of stingless bees with mass provision exceeds the development times of *Apis mellifera* by far. Whereas the honey bee worker emerges from her cell 20 days after oviposition, this time ranges between 45 and 55 days for stingless bees, which seems to be a substantial advantage at first sight. However, in phylogenetically more closely related bumblebee species, it also only takes approximately 25–34 days from oviposition to adult bee (Pereboom, Velthuis, and Duchateau 2003). Is that average 7-day difference sufficient of an advantage to evolve a completely novel feeding regime for larvae?

For the individual larvae, it may only be a 30% difference in developmental time, but at the colony level such a difference may actually make a major difference. The queen can lay approximately 2000 eggs per day, and this accumulates to an impressive 14,000 workers within these 7 days, which is approximately one-third of the worker force in a colony. When there is a rich nectar flow, the average workers are very short-lived (on average, 15 days) and need to be replaced at a high rate to maintain colony strength. It is one of the many examples in which a seemingly small difference at the individual level can have a massive impact at the colony level. The short developmental time for workers is one of the major drivers of huge adaptive flexibility at the colony level. A small colony coming out of winter with approximately 2,000 workers can grow within 4 weeks by an order of magnitude if conditions are favorable. It is then ready to exploit any major nectar flow.

In addition to swift worker development, queen developmental time should also be as short as possible. Replacing the queen is essential for the colony to survive as an entity, and we discuss this in great detail when dealing with the queen rearing paradox (Chapter 5). However, this leaves us with an open question: Why not have the workers develop at a speed similar to that of the queen? If brood developmental time is so important for colony fitness, this would allow the colony to respond even faster to short-term foraging opportunities. It may well be that trade-offs with regard to the bauplan and developmental constraints set for caste differentiation—clearly a major central element to colony functioning—preclude the shortening of the worker development period further. Nevertheless, this shows that the solution that has been achieved biologically may not be the optimal developmental speed for the worker caste. Using a suboptimal diet for larvae destined to be workers is obviously the tool used to ensure that a worker caste is produced. This seems to be much more important than shortening the developmental period at the risk of rearing queen-like workers that might generate conflict in the colony.

The male brood is fed the worker diet, and their developmental time is even longer. They become bigger than workers because they develop in larger brood cells. Because they receive the same food, they need to be fed for longer to reach the larger body size. At this point, we do not further explore whether this is a clever trick on the part of the honey bees. We return to this topic when dealing with sex and reproduction (Chapter 8). For the moment, it seems clear that size matters when it comes to the male sex in honey bees.

{ 4 }

The Chemistry of Social Regulation

FIGURE 4.1 *Workers engaged in retinue behavior released by the pheromones of the queen's mandibular glands.*

Releasing chemicals into the environment is the most ancient form of coordinating the activities of different individuals. Odors can attract and repel based on their chemistry and concentrations, and they are effective tools for forming or disrupting the cohesion of a pair or a group (Figure 4.1). Even in human cultures, chemical signals are often used as perfumes to attract individuals of the opposite (and, less frequently, the same) sex. Such odors need not be acquired in the perfume shop. The peptide ligands of the major histocompatibility complex have been found to be important for assessing the odor of a member of the opposite sex (Milinski and Wedekind 2001; Milinski et al. 2013). Sometimes, however, a specific smell may not be as convincing as the user had intended and actually causes repellence rather than attraction, depending on individual preferences and experiences. Hence, we say that successful groups "have the right chemistry," which is both metaphorically and literally true.

If the right odor composition is important in some way to the formation of groups and pairs in humans, its importance is much greater in most insects that do not have other means of communication to mediate their responses. Their odors are not just "perfumes" in the human sense but form the very basis of information exchange: They are the equivalent of words in a sentence. This is certainly true for social insects. They use chemical signals to regulate the behavior of individuals in ways that are important to the context of the signaler. This ranges from the production of highly volatile, transient signals that are used in raising alarm responses and repelling unwelcome attention from other organisms to the use of compounds of low volatility as sex attractants and primers of changes in physiology with their associated behavioral changes. This diverse array of pheromonal compounds has become integrated into the arsenal of potential signals that social insects use to regulate the behavior of individuals in the group (Figure 4.1). Without pheromones, there would be no social insects, but are these chemicals always used in an efficient way to make a colony work in a well-organized manner?

In the social insects, this transition to chemical communication is achieved through the use of pheromones—chemical signals that demonstrate the presence of the dominant and ensure that her subordinates are prevented from reproducing. This is clearly demonstrated in some species of ants, in which the alpha individual produces a chemical that signals its dominance (Figure 4.2; Peeters, Monnin, and Malosse 1999). What is key here is that there is both a "symbolic" element in signaling presence and a physiological effect on the ovaries and behavior of the subordinates.

Biologists are divided in their opinions regarding the nature of this signaling: Does it represent an honest signal that workers respond to voluntarily or are workers controlled by the signals (Le Conte and Hefetz 2008)? Voluntary

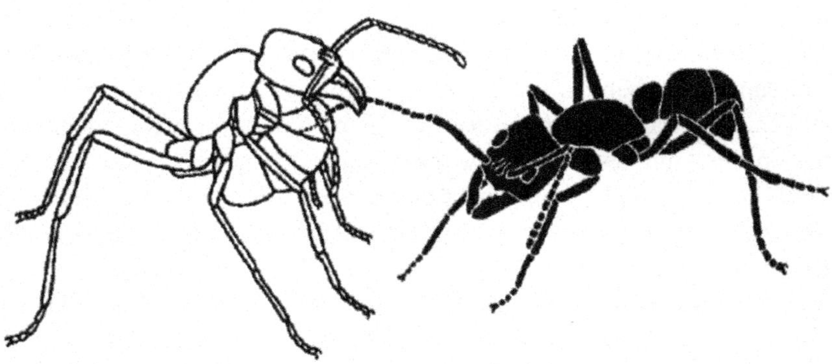

FIGURE 4.2 *The dominant alpha individual on the left exposing the antenna of a subordinate to the secretions from her abdominal glands.*

acquiescence is considered to be the more likely of the alternatives because the claim is that the workers enhance their inclusive fitness by following this behavioral route, and it is considered to be an evolutionarily more stable strategy. Worker control is considered to be less likely since it is thought that it will lead to an arms race between queens and workers. This unresolved controversy remains in place as distinguishing experimentally between the alternatives is difficult (Kocher and Grozinger 2011). As social groups become larger, colony-level selection will become more important than individual selection on queens and workers. Overt conflict in the hive will reduce its efficiency. Because these different levels of selection may not work in concert, it is difficult to distinguish between the alternative explanations for the queen–worker interactions.

4.1. Nonvolatile sex attractants

Just as the origin of social bees is shrouded in the mystery of an inadequate fossil record and contradictory signals from extant species, so the choice of chemicals that are in use in communication by the social insects is equally poorly understood from a phylogenetic standpoint. Whereas a suite of pheromones has been chemically identified and their function has been unraveled, the question as to *why* particular compounds are used for signaling specific information has rarely been addressed. Clearly, volatility plays a role based on the function of the signals, but also the evolutionary history of the organism determines *what* specific compounds are used.

The ancestral form of chemical communication in eukaryotes is the use of specific compounds to attract members of the opposite sex. The majority of insects use pheromones like we use perfumes: to attract mates. Sexual selection may have driven the co-evolution of specific chemical signals from the signaling individuals and specific detecting apparatus in the receivers. The seminal example in insects is that of moth sex pheromones described so eloquently by Fabré in his *Souvenirs Entomologiques* (1879–1907/2002): Describing male peacock moths attracted to a newly emerged female confined in a cage in his study, he wrote, "Forty lovers eager to pay their respects to the marriageable bride born that morning amid the mysteries of my study" (Fabré 2002, p. 249). This work established that female sex pheromones were disseminated as an odor plume that attracted males from a significant distance. The details would have to wait for chemical analysis to catch up approximately 50 years later with the identification of bombykol, the first sex pheromone characterized in the silk moth by Butenandt et al. (1959).

Characterization of the insect sex pheromone communication system has shown that the compounds constituting the pheromones are of relatively low volatility. The receptors that detect these compounds on the antennae need to be exceptionally sensitive, allowing for responses to individual molecules. Male silk moths respond to single bombykol molecules in search of females. In bees, the females produce sex pheromones that attract males for mating because the males are active seekers after females (Paxton 2005). Honey bees follow this pattern in having the females produce fatty acids as the sex attractant for the males. These signals, like bombykol, have low volatility. This does seem to be highly maladaptive at first sight. Shouldn't these signals be volatile like the perfumes we use? The vapor pressures of the fatty acids that are used as the sex attractant are three or four orders of magnitude lower than that of volatile perfumes. So clearly these chemicals cannot be perfumes in the human sense. Compounds of low volatility do not decay over distance as quickly as the perfumes with high volatility. As with the male silk moth, the drones evolved extremely sensitive receptors for these fatty acids to detect the queen on mating flights (see Chapter 8). It is the low volatility of the compound that allows the drone to perceive the queen signal over long distances in flight far away from the colony. If it were volatile, the signal would spread broadly in many directions, making it difficult for the drone to use it as a guiding signal.

4.2. The struggle for position in reproductive hierarchies

For the queen, attracting a few males to mate with her is a relatively simple task by comparison with organizing the other 40,000 females in the hive. Can she use the same pheromone to attract males and to prevent reproduction by the host of workers in her colony?

If we are to understand the ways in which the queen rules through the use of efficient chemical communication, it is essential to explore the evolutionary changes to the chemical signals used by honey bees' solitary ancestors. Compounds that are ideal for attracting a sex partner of a solitary bee may no longer be suitable for a highly social insect. Yet evolution may still take advantage of these ancestral chemicals and use them in the social context because they are already present in the solitary bees. Pheromonal signals may be chemically conserved, whereas audiences may change as social complexity increases.

Long-lived eusocial colonies provide for the emergence of a set of audiences that respond to both physical and chemical signals differently. Both the audiences that are addressed and the stage on which the play unfolds are essential for the messages to be communicated. Just imagine hearing your favorite dance music in an airport lounge as opposed to your favorite disco club. The signal will release fundamentally different behaviors depending on location. At the same time, the signal may attract some people into the lounge, whereas others will be happy to reach their gate half an hour earlier. So the same signal may mean different things to different sets of individuals. Yet is such variability in signal response adaptive in the honey bee colony? Wouldn't it be better to have a clear signal for one group and a different one for another so that there is no confusion in the hive?

The key factor in the development of harmonious insect societies is that some individuals are forced to give up reproduction in favor of raising the offspring of relatives. This social organization is based on a radical skewing of the reproductive potential of females so that one member of the group reproduces and others assist, which seems reasonably unfair and certainly undemocratic and creates a situation in which individual fitness is lost. As these group-living bee species evolve, individuals contend with each other for dominance, and the one who wins physically becomes the reproductive whereas the others become effectively sterilized helpers. There is always an uneasy set of relationships within the group that may be upset if the dominant individual is removed, weakens, or dies, in which case other members of the group will attempt to become the monopolizers of reproduction.

Physical interactions allow for a span of control within groups that is limited by the number of individuals that can be effectively regulated by direct physical interaction. As groups increase in size and complexity, there is a necessity to develop alternatives to direct interaction between all individuals in the group if it is to remain coherent. The dominant individual needs to advertise its presence through alternative channels of communication so that it reaches a larger audience than is possible through direct physical contact.

In the phylogenetic ancestors of the honey bees with less developed levels of sociality, groups usually consisted of a mother and her offspring that were closely related or a group of offspring. The allodapine bee species (Figure 4.3) provide good examples of the way that social transitions may mold individual interactions, with some evidence that pheromonal means of regulating female reproductivity (O'Keefe and Schwarz 1990) arise even in small groups (<10 individuals) and can serve as the substrate for evolutionary development of more complexity as castes and large colonies come into existence. Caste emerges in the allodapines as a result of behavioral differentiation between individuals that specialize in different tasks (Schwarz et al. 2011). As eusociality emerges in some species, caste is not

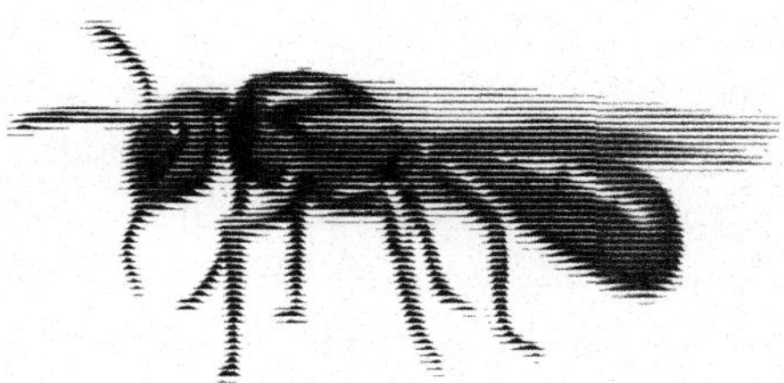

FIGURE 4.3 *Exoneura bicolor bees nest in small communities with progressive feeding of the larva.*

only characterized by differences in behavior but also acquires increasing differences in morphology. This characteristic of morphological differentiation is a key element in the evolution of the highly eusocial honey bee colonies with the establishment of a morphological worker caste of functionally distinct individuals.

Assuming that chemical signals of queens evolved from those used in the ancestors of the eusocial bees, we need to explore what has emerged from this evolutionary process. Some common features of the chemical precursors of pheromones are becoming clearer from the large numbers of compounds that have been identified in insects. The chemistry of most insect pheromone compounds is based on the use of long-chain fatty acids (LCFAs) as the starting material from which the pheromone molecules are made. These LCFAs form part of the ubiquitous phospholipids used for intracellular lipid storage and membrane synthesis (Stanley-Samuelson et al. 1988). The molecular machinery of the cell allows for the lengthening of the fatty acids from the standard lengths of C12–C18 up to waxes of chain length C40. The insects use these long-chain compounds to synthesize cuticular hydrocarbons (CHCs) that cover the cuticle, and in the honey bees some cuticular cells have been modified to produce the wax from which combs are constructed. Similarly, pheromones can be produced through the chemical modification of fatty acid functional groups and through chain shortening to produce compounds with a range of volatilities suited to the context in which the signal is being used. In the case of the Lepidoptera, chain lengths vary from C10 to C21, and most of the compounds are not free fatty acids. Female bees often use cuticular hydrocarbons as sex pheromones or as a marker of individual identity (Ayasse, Paxton, and Tengö 2001). Although there is evidence that long-chain CHCs may be evolutionarily conserved queen pheromones, the honey bee predictably does not fit this mold (Oi et al. 2015; Van Oystaeyen et al. 2014).

4.3. From a social pheromone to a sex attractant?

If the hypothesis that long-chain hydrocarbons are ancestral queen pheromones is correct, then the immediate ancestors of the honey bees made a switch to using a C10 fatty acid as the basis for the signal that would attract mates. Rather than producing it from abdominal tergites or sternites, they produce it in the mandibular glands. The biosynthesis of this compound has been studied in great detail. The starting point for the synthesis of the fatty acids is the common fatty acid, stearic acid (Box. 4.1). This acid is hydroxylated to produce the fatty acid alcohols with the hydroxyl group either at the end of the molecule or one carbon atom in from the end of the molecule. These fatty acid alcohols are then chain shortened to the C10 fatty acids—10-hydroxydecenoic acid (10-HDA) or 9-hydroxydecenoic acid (9-HDA).

BOX 4.1 The Chemistry of Social Regulation

A bifurcated biosynthetic pathway from stearic acid, a C18 fatty acid, differentiates the products of the mandibular glands of workers and queens of honey bees into the major components 10-hydroxydecenoic acid (10-HDA) and 9-oxodecenoic acid (9-ODA) respectively. A specific hydroxylation step adds the hydroxyl group either at the end of the stearic acid molecule (ω hydroxylation) or one carbon atom in from the end (ω-1 hydroxylation) and is the first essential point of differentiation in the pathway that produces caste-specific products. Chain shortening from C18 to C10 and the addition of a double bond between C2 and C3 are common in both pathways. The final oxidation step that produces 9-ODA, the "queen substance," is the second essential point of differentiation. The queens use audience and location differentiation to use the same suite of chemicals for different purposes: releasing mating behavior (audience of males) or informing the colony (audience of workers) about her presence and affecting the physiology of the members of this audience. The workers use 10-HDA, the "worker substance," as an antimicrobial compound. It is also important in the regulation of the viscosity of the larval food (see Chapter 5).

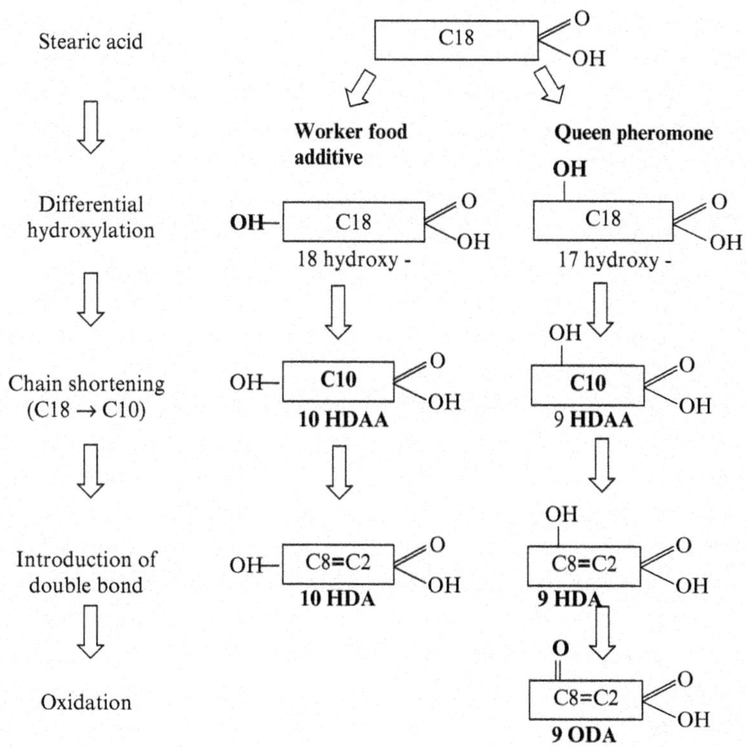

So why would honey bees not use cuticular hydrocarbons as a sex pheromone but use fatty acids instead? Why develop an entirely new sex attractant? It may well have to do with progressive feeding and the peculiar diet of the larvae. The peculiar feeding regime of the larva is based on food gland secretions produced by workers. As discussed in Chapter 3, the larval food is a proteinaceous and sugar-rich secretion produced by the hypopharyngeal glands. This food is deposited in the bottom of the cell for the larva to eat at a temperature of 35°C and 60% relative humidity, which are ideal conditions for bacterial growth. It is therefore not surprising that the bees add antimicrobial compounds to prevent fermentation. Haydak (1970) first recognized the potent antibiotic potential of 10-HDA, which is secreted by the mandibular glands of workers and added to the larval food. This, together with nest hygiene, is particularly important given the large brood nest and the successive use of cells to rear larvae.

In this scenario, the biosynthetic machinery that produces an important antibiotic component of larval food in the mandibular glands was in place and essential for producing offspring. However, whereas the nurse bees are busy producing antimicrobial fatty acids for the conservation of the larval food, the queen does not feed the larvae and she has no need to produce any antimicrobial food additives. She is fed these compounds by her workers but nevertheless has the glands that are perfectly suited to synthesize fatty acids. Rather than shutting down the synthesis, she apparently modified the 10-HDA biosynthetic pathway in her mandibular glands to produce 9-oxodecenoic acid (9-ODA), which has not been shown to have an antibiotic function. Before it became a pheromone, she probably produced it only because she could. What else to do with these funny glands? However, because she is the only female who produces this compound, it of course has all the characteristics of a suitable signal for the workers to use to perceive the presence of the queen in the colony—a necessary condition for any queen pheromone. Since 9-ODA arose as a queen-specific pheromone with low volatility, it is perfectly suited to attract males in flight. Thus, the novelty of the honey bee choice of pheromones is that in using C10 free fatty acids, they have compounds with relatively low volatility (comparable to that of C20 lepidopteran sex pheromones), which is effective in the confines of colony as a social regulator (comparable to the CHCs used by other social insects) and is also effective in mate attraction. What seems at first sight to be an odd anomaly turns out to be a rather efficient means of communicating to both workers and males. Whereas the queen modifies the mandibular gland fatty acid biosynthetic pathway to produce the queen substance 9-ODA, the workers retain the original antibiotic function of the 10-HDA fatty acid, which is a key component of the larval food (see Box 4.1). In this scenario, the antibiotic component of larval food is modified to become a social pheromone, which is then co-opted for use as a sex attractant. This reverses the evolutionary direction suggested by the long-chain CHC hypothesis for the evolution of these pheromones.

{ 5 }

The Reproductive Machine

FIGURE 5.1 *A queen of Apis mellifera.*

The only fertilized fertile female in the colony has been termed the queen (Figure 5.1). Calling the single fertilized reproductive female a "queen" draws us immediately into a conceptual controversy arising from the comparison of human and insect societies. No lesser person than Aristotle (350 BCE) started this rot when he claimed that "working bees are engendered by the rulers of the hive" and contributed to the false image of a royal ruler of the hive. The queen, one of the better studied individuals in the colony, has none of the attributes of real human queens, which is why students of entomology prefer to call her a *gyne*. However, in apiculture the honey bee gyne has been termed "queen" in all languages that we are aware of, and so we believe we should stick to this term throughout this volume: The honey bee gyne is the queen (and why not, since even the group of outstanding musicians associated with Freddy Mercury in the 1980s chose this equally odd term for the name of their band, most likely not thinking about honey bees at all). In brief, the honey bee queen is on average two cm long, weighs 190 mg, lives for up to four years, and can lay up to 2,000 eggs a day (for details, see Winston 1987). She has a spermatheca to store semen after mating, and once her ovaries are activated, she will not mate again (see Chapter 8). The queen cannot collect any pollen or honey and is fed throughout her life by the workers with food jelly (see Chapter 3). This high-protein diet allows her to maintain her extreme egg-laying capacity. So all in all, her fecundity seems to be perfectly designed for the job of maintaining a large, viable worker force in the colony.

5.1. Caste determination is sweet

Queens are derived from fertilized eggs, and the larvae destined for this reproductive role are fed a special diet, the royal jelly secreted by the nurse bees. Royal jelly seems to taste good, and the royal larvae receive and eat six times more of it than do worker larvae (Haydak 1970). The rich diet makes queen larvae grow large within a short period of time and requires that they be reared in enlarged brood cells to accommodate their greater size.

For many decades, the search for a queen determinator in royal jelly that would ultimately define caste was a major research quest (Remboldt and Hanser 1964; Remboldt, Lackner, and Geistbeck 1974; Remboldt and Lackner 1981). Despite this strenuous quest, the search for this key substance has remained elusive because it could not be clearly identified. Eventually, Asencot and Lensky (1976, 1984, 1985) came to the conclusion that sugar concentration was the prime driver of caste determination. They added sugar to the food jelly in worker cells and were able to produce queen larvae. Most students in the field of social insect research were satisfied with this explanation and believed this had plausibly resolved the issue. The search for a determinator was over, and those larvae that ate a lot of sweet-tasting sugar-rich food grew big and strong and eventually became queens, whereas those kept on a limited low-quality diet developed into workers. Today, it is known that DNA methylation and epigenetic mechanisms are essential for differential activation of gene cascades that lead to a queen rather than a worker. Nevertheless, Kamakura (2011) reopened this issue by publishing a report in *Nature* magazine reviving the determinator concept in the form of royalactin, a royal jelly protein that had been found to function as a growth factor in the fruit fly *Drosophila melanogaster*. He reported that female honey bee larvae reared on freshly harvested royal jelly developed into queens, whereas larvae reared on royal jelly stored for 40 days (after which the royalactin present had been degraded) developed into workers. Adding biotechnologically produced or purified royalactin to this degraded royal jelly supposedly restored the development of queens in his experiments. When royalactin was fed to *Drosophila* fly larvae, they developed into bigger individuals, and because the effects of royalactin could be inhibited by knock down of the epidermal growth factor receptor (EGFR), Kamakura concluded that the protein factors in the larval diet interact with EGFR to eventually determine queen caste. However, it was a pity that these results were not discussed in the context of the sugar concentration hypothesis. In fact, Asencot and Lensky (1988) also compared stored versus fresh royal jelly and obtained a similar result but came to a very different conclusion. They found that it was the content of soluble sugars, but not crystalline sugar, that contributed to caste differentiation. Because Kamakura also added water by adding the protein (and given that he followed his protocol, he was very likely only adding water because he discarded the majority of the protein fraction since it was

FIGURE 5.2 *The sting of a worker (left), intercaste (center), and a queen (right). Whereas the queen has a curved sting shaped to kill its rivals, the worker sting is straight with barbs. The intercaste sting has no barbs and is slightly curved.*

not water soluble), crystalline sugars in the aged royal jelly may have been redissolved and triggered the queen developmental pathway. Hence, the evidence for royalactin as a determinator compound remains elusive at best if the effects of sugar concentration are not rigorously controlled (Buttstedt, Moritz, and Erler 2014). Showing that honey bee proteins can make big flies is correlational evidence confirming that peptides in the larval diet of honey bees can influence insect growth. This is an important finding, but it does not provide evidence of the existence of a "determinator" in honey bee caste development. The differentiation of castes is not just about growth; it is also about developing a completely different phenotype. It is about releasing the full potential of the phenotypic plasticity present in the honey bee's genome. The queen is a completely different organism from a worker, not just bigger. One needs to explain the allometric growth that results in a fundamentally different morphology and physiology of a queen compared to that of workers (Figure 5.2). Buttstedt et al. (2016) showed that degraded royal jelly even after proteinase K treatment was perfectly suitable for rearing queens in vitro. Indeed, adding royalactin to a degraded diet had no effect on the caste fate of the larvae. So, yes, experiments have been conducted that definitively reject the determinator concept. Unfortunately, the publication of Kamakura did not refer to the older studies and received high visibility at a time when the question of the existence of a determinator had already been settled.

5.2. The role of proteins in royal jelly

Are these proteins just providing a balanced diet to a tasty solution of sugars? Most likely not. Indeed, the proteins have important functions that are neither related to diet nor to caste determination as suggested by Kamakura (2011) because their presence controls the consistency of the larval food. Most important in this context is the pH of royal jelly. In addition to the secretions of the food glands, the workers also add fatty acids secreted by the mandibular glands that makes the food sour (pH = 4) and profoundly affects its consistency. We all know this from souring milk where the proteins precipitate. This process can be easily reversed by increasing the pH, which changes royal jelly from a solid ivory-colored gel to a watery transparent liquid. Buttstedt et al. 2018 showed that two proteins, major royal jelly protein 1 (the oligmeric form of royalactin) and apisimin are required to keep the viscosity of the food jelly at the correct level so that the larvae can take up the food in the cell. They float on the viscous jelly surface so that they have access to the food but do not drown. Changing the pH of royal jelly only slightly from pH 4 to pH 5 with alkali causes queen larvae to fall out of their cells because the viscosity of the food has changed (Figure 5.3). Without the high viscosity of royal jelly, rearing queens would simply be impossible in vertical cells opening downward. Indeed, major royal jelly protein 1 (royalactin) occurs in various forms. The monomers readily combine to form oligomeric structures, with the level of oligomerization increasing with declining pH. These oligomers together with apisimin eventually start building large networks of protein fibers that increase the viscosity of royal jelly.

In addition to these physical properties that are essential for successful brood rearing, some proteins and peptides in royal jelly are known to have potent antimicrobial effects. For example, royalisin, a honey bee defensin and highly efficient antimicrobial peptide, is a component of royal jelly. In addition, major royal jelly protein 1, although not the queen determinator, has

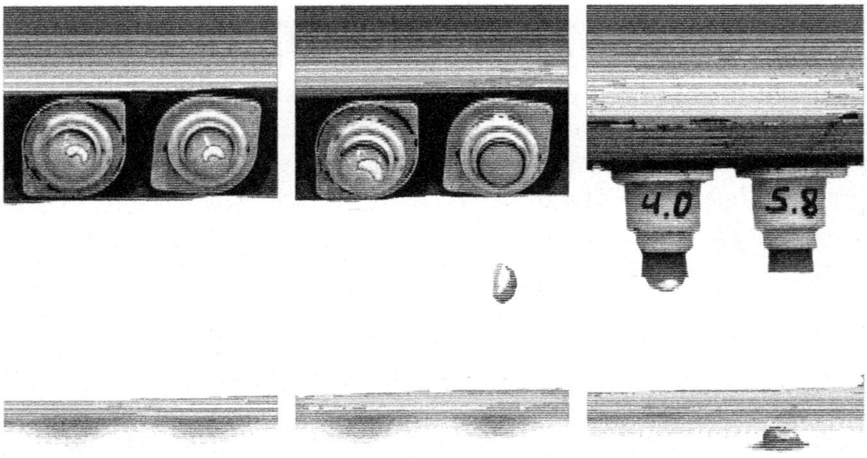

FIGURE 5.3 *Queen larvae placed on royal jelly in standard artificial queen cells used for queen rearing in apiculture at pH 4 (left) and pH 5.8 (right). After transferring a larva into each of the cells, the frames were then slowly rotated (left) into the vertical position (right) found in a hive. Whereas the queen larva in the left cell adhered to the royal jelly, those on the royal jelly at pH 5.8 fell out onto the floor (center and right).*

strong antimicrobial activity against various honey bee pathogens (Buttstedt et al. 2016).

In summary, the proteins in royal jelly are clearly more than just food providing the right mix of amino acids. They are essential for keeping the food gland secretions in a jelly-like form and have antimicrobial properties as well. In addition, the low pH is essential because it also has antimicrobial effects and keeps the proteins stable over time.

5.3. The etiquette of queenly interactions

Once a queen-destined larva has become an adult queen, she is the sole reproductive (most of the time) of the colony and hence functionally in a most central and important position. Indeed, she has a strong regulatory function that provides colony cohesion which is primarily driven by a complex set of chemical signals. A suite of glands produce a variety of signals that are used by the queen to communicate with the other members of the colony. These can attract, repel, release aggression, or appease. If the colony were hierarchically organized, like a large army, the queen could in principle simply release the appropriate signal at any given time such that the colony would behave in a coordinated adaptive way to the liking of the queen. Fortunately, such a system did not evolve, and indeed if it had ever been in the infinite toolbox of evolution, natural selection would have immediately eliminated such complete queen control.

What does the queen know about the outside world? Next to nothing, which is probably very similar to royal leaders of the past in human societies. The workers engage with the environment and find food, water, and eventually new nesting sites (see Chapter 6). The day-to-day activities of the colony are regulated in a decentralized manner without the need for specific pheromones released by the queen. The queen has one main purpose: to be the only female producing offspring. It is here that her chemical signaling becomes important to control worker reproduction, as we illustrate in detail in Chapter 6. Yet her signals are not just important for the worker audience. The queen's pheromones assist her in maintaining her relationships with a set of audiences that appear and disappear as her life unfolds. This starts immediately on emergence from her cell. In this phase, she actually conceals her presence from workers by not producing the royal signal in the colony. She thus remains unnoticed by most other members of the colony as she has to resolve a major problem: another queen.

She will become the only fecund female and hence she needs to find all rival queens within the crowd of workers and kill them (Figure 5.4). If she is the first to emerge, she will attempt to detect the presence of her queen sisters in their cells and kill them. Curiously, these sister queens advertise their presence

FIGURE 5.4 *The lethal fighting behavior of virgin queens is released by signals from the tergal glands.*

by producing a piping sound that may have the effect of ensuring that they are killed by their siblicidal sister. If this is true, it certainly does not represent a particular adaptive behavior to enhance their individual fitness. Shouldn't they rather sit still and be as inconspicuous as possible?

If the first queen that emerges fails to kill its sister queens and other queens are released from their cells, then the scene is set for dramatic virgin queen duels. Now chemical signals come into play. The recognition and detection of rival queens is done via highly specific pheromones released in the tergal glands. These glands secrete a mixture of oleic acids (hexadecyl linoleate, hexadecyl oleate, and octadecyl palmitoleate), which immediately induce the lethal duels between queens that reduce queen numbers to a single survivor. In this process, workers may attempt to prevent the duels between the competing queens but have their efforts rebuffed by virgin queen defecation in which a specific repellent component (*O*-aminoacetophenone) is released. This fecal component keeps the workers at bay while the virgin queens continue their dance of death. The defecation pheromone reduces the workers' ability to exercise any choice in relation to the virgin that ultimately survives to have an opportunity to reproduce. On most occasions, only one virgin queen prevails and is now ready to meet with a new audience—the drones outside the hive at the drone rendezvous areas.

5.4. Meeting the drones

After the queen survives her duels, she obviously needs to mate with males—and many of them. She has to produce some millions of fertilized eggs in her lifetime and needs to collect semen from many males during a sequence of several mating flights. She stores this semen in the spermatheca, which has to last her entire lifetime as she will never mate again after these flights. It would be highly maladaptive to mate with her brother drones inside the safety of the hive because this would result in extreme inbreeding and be lethal for the production of offspring due to the peculiarities of the honey bee's sex determination system (see Chapter 8). The virgin queen therefore embarks on extensive mating flights in search of drones. Yet the area in which mating can take place is large and colony densities may be low. So how to attract drones in mid-air far away from the hive? This is hardcore ancestral pheromonal sex attraction biology. The queens produce compounds of low volatility (primarily the queen substance, 9-oxodecenoic acid (9-ODA)) that are released in flight and are attractive to drones at rendezvous areas termed "drone congregation areas." Drones from a number of colonies are attracted to these locations in search of queens. The pheromone components are released into the airstream as the queen flies, and the males detect the chemical and orientate to the queen by flying upwind toward her. The male antennae have highly specific olfactory apparatus and a means of processing the olfactory information that is exquisitely finely tuned to detecting the queen pheromones. The drone olfactory system is not designed to detect a range of potential perfume sources but, rather, is highly selective for a particular pheromone cue at low concentrations. In this case, the stage on which communication takes place is in flight, and the audience is exclusively composed of drones. The drones are designed to be attentive to the young queen's sex pheromone because their olfactory system is designed to detect very low quantities of these chemicals. In this case, the release of the pheromone (9-ODA) that has low volatility (1.77×10^{-6} mmHg) is by diffusion and then mass transfer into the airspace behind the flying queen. It is this plume of pheromone molecules that the drones detect and then use to orientate toward the flying queen. Jean Christophe Sandoz (2006) has visualized the exquisitely finely tuned olfactory system of drones by studying the neuronal response in the antennal lobes of the brain (Figure 5.5).

As more drones detect this odor trail, their aggregation becomes a visual cue for other drones to join the group and attempt to follow the receptive female. This signaling system is highly effective in allowing queens to be mated multiply in a short period of time. In addition, the signal that the queen produces is dispersed in the atmosphere by turbulent flow of air. Once the queen leaves, the signal rapidly disappears.

Once a drone has physically located a queen, he orientates to her visually, and it may be that the queen's tergal gland secretions assist the male with capturing her through short-range orientation cues. This highly polyandrous mating system will ultimately provide the reproductive female with a colony of daughters with a diversity of genetic backgrounds determined by the number of males that successfully mate with her and whose sperm can be used to fertilize her eggs (Figure 5.6). The mating behavior of honey bees is curious in that it takes place at a distance from the colony and exposes the reproductive females to high risks associated with flight outside the nest (see Chapter 8).

Having now acquired the sperm that she needs for the rest of her life, the young queen concludes her "discussion" with the audience of drones outside

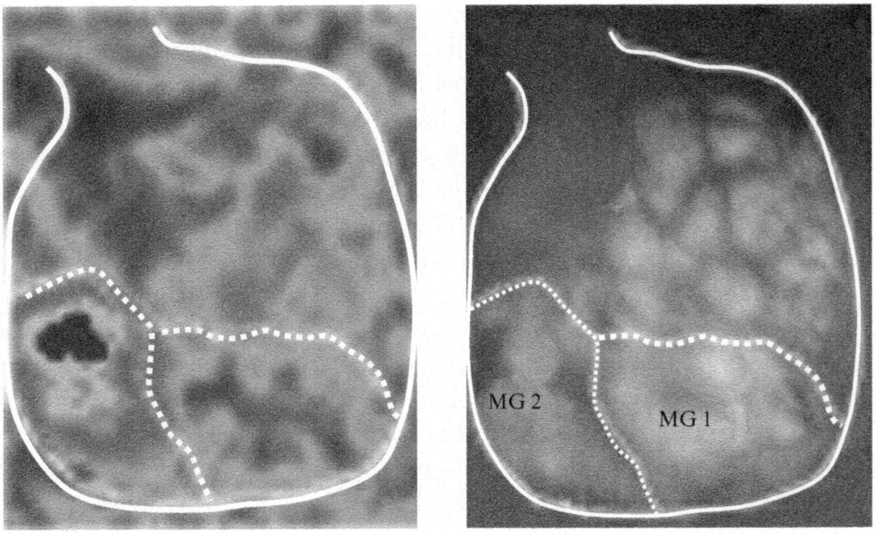

FIGURE 5.5 *The left antennal lobe (in gray image) in the brain of a honey bee drone after exposure to 9-ODA. Only a particular region in macroglomerulus 2 (MG 2) responds to the pheromonal signal visualized by calcium imaging (left, black spot on image. For color version see tuck-in page at end of the book).*

FIGURE 5.6 *Drone just before mating in flight with a queen and his subsequent instantaneous death as he pumps all his hemolymph into his endophallus.*

the colony and returns to the colony to start the next phase of her life with the audiences within it. However, back inside the hive, the queen's perfume presents a serious problem for the drones within the colony. They are now exposed to an environment that is pervaded by the sex pheromone at very high concentrations. Yet despite their exquisitely attuned olfactory system, they do not respond in the confines of the colony. Why don't they mate with the queen right in the colony if she releases such a powerful sex attractant? We can speculate that their olfactory receptors are saturated inside the colony and that they are habituated to the queen's sex pheromone and hence do not respond. However, how do they manage to reverse this saturation when they leave the colony for a mating flight? Obviously, they cannot magically reset the olfactory system to a highly sensitive position so they can be attracted to the queen substance during a mating flight. The mechanism whereby drones achieve the sensitivity toward the queen substance from one minute to the next is completely unknown. It may well be that during the flight itself the pheromone molecules are removed from the receptors. Maybe this is a major reason why mating flights typically last longer than 30 minutes. Only after a sufficiently long period of flight when the queen substance has been removed from the sensillae may they be able to orientate to queens at the rendezvous areas and be an attentive audience to the siren perfume of the virgin queens. Whatever the mechanism, clearly the use of a specific sex pheromone by the virgin queen would have made it much simpler for the poor males to spot the queen. They could orient toward a pheromone signal immediately after embarking on a flight, probably reaching queens much faster and hence with a lower risk of missing out on a mating opportunity. Thus, the dual function of the queen substance as an in-hive pheromone and an outside sex attractant does not make it particularly easy for the males to orientate to queens in the colony but does ensure that they can find queens outside the colony once their olfactory systems have been reset.

5.5. The problem of an inattentive worker audience

So what is the fate of the perfume used to lure males once the queen has returned to the hive? The chemical signal used for mate attraction has lost its audience (males) because mating has been consummated and no further matings will take place during the lifetime of the queen. It is now time to produce the brood and ask the workers to rear the eggs to adults. So wouldn't it be efficient to use the same compound that has been used to attract the males to inform the workers that there is a mated queen in town? However, communication is not easy. The workers lack the highly sensitive receptors of the males. Single molecules are insufficient to inform them about the presence of the queen. They apparently need a lot of signal to recognize the presence of the queen and remain sterile, and they are rather insensitive compared to the highly tuned drones on their mating flights. The queen needs to ensure a constant flow of high-dose pheromone, which requires quite some effort from her. She produces between 20 and 500 μg of queen substance per day to keep the self-organized feedback loops controlling the fatty acid pathway in workers from switching to the queen pathway. As soon as queen substance levels become low within the colony, the workers switch their own pheromone production to the queen pathway and produce it themselves. This again establishes pheromonal hierarchies in the colony, with the majority of workers retaining the worker pathway in response to the queen substance produced by the few dominant workers. Those producing more queen substance in turn become less sensitive to its suppressive effects, which allows them to further enhance its synthesis and eventually results in the development of worker false-queens, to which we return when discussing the dark sides of worker behavior.

So the queen needs to signal her presence and reproductive dominance to ensure that none of the workers over which she has hegemony start reproducing. The initial stages of this hegemony are established by rapidly increasing the level of production of secretions from the mandibular gland and augmenting this at a later stage by secretions from the Dufour's gland (long-chain esters and hydrocarbons) and those of the tergal glands. The composition of the secretions from these three glands changes as the queen establishes her position as the dominant egg layer and captures the exclusive attention of her audience of non-reproductive workers. This conversation between the queen and workers proceeds smoothly as long as the queen is attended by a court of workers who receive her queen mandibular gland pheromone and distribute it throughout the colony so that the entire workforce is aware of her presence.

However, the colony faces a problem. Workers need to be informed quickly of any changes in the condition of the queen and hence the turnover of the pheromone needs to be efficient in order to maintain its effectiveness as a signal. Yet the signal is of exceptionally low volatility, so how can workers far away from the queen know she is gone? The distribution of the pheromone is achieved by contact with the queen. The workers that form the court around the queen assiduously lick and antennate her, and through these contacts they become contaminated with small amounts of the queen pheromone (see Figure 4.1).

Queen pheromone-tagged workers then become "messenger" bees that leave the court and carry the message of the presence of the queen to the far reaches of the hive. The signal is throughout the hive, so how do the bees know the queen is no longer present? Because the pheromone is of low volatility, it should actually be a persistent contaminant of the hive interior, and it should be completely unsuitable for indicating the absence or presence of the queen. Indeed, evolution had to design a way out of this pickle of using a stable signal of low volatility instead of a volatile, rapidly decaying signal in case of queen loss. The main pathway used to get rid of the pheromone and ensure a high turnover appears to be simple. The workers simply ingest the compound. They actively destroy the queen pheromone by the process of internalization, which probably results in its metabolic destruction. The result of these processes is that the absence of the queen can be detected within a period of approximately 20 minutes despite the low volatility of the signal (Naumann et al. 1991). In addition, the fatty acids of the queen signal are lost in comb wax and diluted through distribution to increasing numbers of recipient workers as the messenger bees wander through the hive. Hence, the

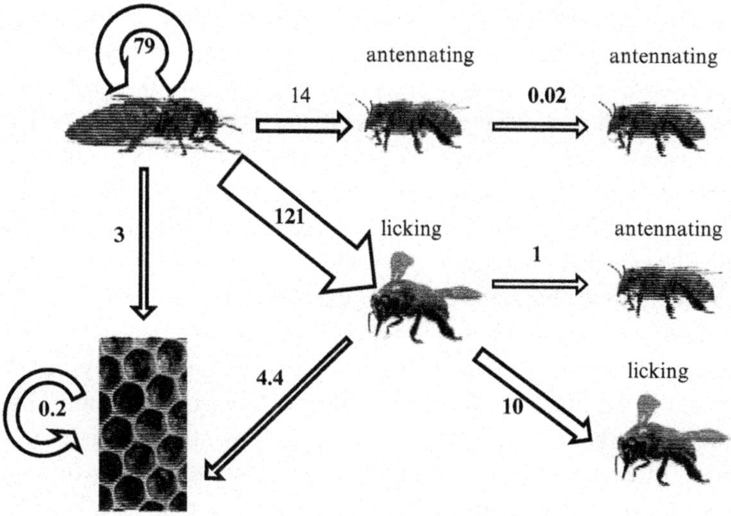

FIGURE 5.7 *The distribution and removal of queen substance in the colony with the queen as a source of continuous production was modeled by Kenneth Naumann in his PhD dissertation in 1992. In this example, the queen produces 217 μg of queen substance/day, of which 79 μg is immediately internalized, 121 μg/day is passed on to the cohort of all licking messengers, 14 μg/day is passed on to all antennating messengers, and 3 μg/day is absorbed by comb wax. Smaller amounts are then passed on to other workers that interact with the messengers from the queen's court. Licking workers take up about an order of magnitude higher amount of queen substance compared to antennating workers. The reflexive arrows show internalization of the signal by the queen and the wax. Workers internalize 50% of the signal received.*

equilibrium between pheromone production by the queen and its distribution throughout the colony is highly dynamic. The constant production of large amounts of queen pheromones is essential to successfully advertise the presence of the queen in the hive (Figure 5.7). Should the queen be lost, this dynamic equilibrium starts to decay quite quickly with the removal of the source of production—within a matter of less than one hour there is panic in the hive—and emergency queen rearing starts.

5.6. Panic in the hive

The loss of the fully fecund mother queen causes an abrupt break in the synthesis and distribution of her pheromones that is at the extreme end of the dynamic set of interactions between the queen and her workers. One might expect that any well-adapted social insect colony would have a "spare" queen up their sleeves to immediately replace the old one. Whereas many social insects do so, the honey bees do not. They need to start from scratch, which is why it is necessary to press the panic button in the hive. Activities start immediately after queen loss to rear new queens from the young worker brood produced by the old queen before her loss—a process termed emergency queen rearing. The workers choose specific young worker-destined larvae to feed with royal jelly. The cells in which they occur are expanded, and the royal feeding regime turns these individuals into queens. This is a very critical phase in the life history of the colony because workers tend to shift tasks. For example, queenless colonies are more susceptible to being robbed by other colonies or may be invaded by parasitic swarms that take over the colony.

In African honey bees, the queen replacement phase often takes too long, and a significant proportion of queenless colonies are usurped by parasitic queens or workers. Indeed, where parasitic Cape honey bee workers are present in an apiary, any attempt at queen rearing will fail. Given this risk, the honey bee's strictly monogynous colony structure is not particularly well adapted to deal with these contingencies. Having only a single queen with no backup seems like an unnecessary risk to take. The costs of keeping a spare queen seem to be so minute in relation to its potential benefits. Indeed, it seems that the polygynous social insects are more successful than the monogynous ones. Species with a single dominant female often have mated "backups" that take over when the alpha reproductive fails. Even given that monogyny is somehow important for the honey bee colony, an additional mated queen might not even need to get into conflict if she was pheromonally suppressed by the dominant reproductive. If this works for workers, why should it not work for a subordinate queen? Apparently, the honey bees do not follow the prudent path but gamble on the success of emergency queen rearing. In addition, there are further risks and unnecessary potential conflicts, as discussed later when dealing with the peculiarities of the swarming process.

5.7. In the swarm

Other stages in the life of the queen generate new audiences that need to be engaged and kept attentive. This is particularly the case when the queen is going to leave her colony with a swarm to move to another nesting site. Yet again, it is the queen's mandibular gland pheromones that have been shown to be important for keeping the swarm together as a unit. The swarm can be considered as a three-dimensional expansion of the queen's court from that of a circle of bees around her on the two-dimensional surface of the comb to the whole swarm cluster outside the hive. The queen is at the center of the cluster when the swarm settles before proceeding to its new nest site. If the queen does not participate in swarming (beekeepers prevent this by putting a queen excluder screen at the flight entrance) or is lost in flight, the swarm will lose coherence and eventually return to the natal nest. Also in this instance, the low volatility of the queen's pheromone seems to be a serious problem for keeping the seemingly chaotic cloud of bees together. Is it even possible that a signal with low volatility could function in such a way? Considering the way the pheromone is spread inside the colony, it actually seems rather unlikely that the pheromone is a potential signal that informs the swarm about the presence of the queen. It is clearly impossible to spread the pheromone using messenger bees in flight as is done in the hive. Because all individuals are in flight at the same time, there is no trophallaxis and there is no antennation that would allow for the exchange of surface-based fatty acids. Once the swarm has settled, the pheromone compounds with low volatility may regain functionality as in the hive (Butler, Callow, and Chapman 1964). This will be facilitated by the workers being tightly packed together—even more tightly than in the hive because they all cling together without combs separating them. It would therefore not be surprising to see the queen producing additional, much more volatile signals that inform the swarm about her presence in flight. Indeed, pentadecane and heptadecane have been found to be an order of magnitude more abundant in queens when the swarm lifts off, and these components may provide the volatile cue required to keep the individuals of the swarm together when they are in flight (Richards et al. 2015). Light at the end of a dark tunnel in relation to understanding queen pheromonal communication after all?

5.8. End of the cycle

After the swarm has established itself in a new nesting site, the queen continues signaling her presence to her workers. However, as she ages, the production of her queen mandibular gland pheromone will begin to fall, with a resulting disruption in the dynamic equilibrium in the colony. The queen's discussion with her workers is also influenced by other sets of interactions that go on simultaneously—larvae that are being reared in the colony produce signals that guide the behavior of nurse bees and hence reinforce the effect of the queen's messages. The interests of the queen and larvae coincide in their need to ensure that workers remain sterile and attentive to the work that they are required to do. As long as there is brood to feed, the queen is obviously in good shape. Thus, the use of progressive feeding of larvae indirectly reinforces the regulation that the queen maintains by her own pheromones. The brood produces a signal that contains a mixture of ten fatty acid esters that affects worker behavioral development (Le Conte, Mohammedi, and Robinson 2001), as does a single ester from this mixture, ethyl oleate, produced by foraging workers (Leoncini et al. 2004). All these signals therefore play in concert with the queen to address the worker audience. As soon as one signal falls below a threshold that affects worker behavior, the workers will initiate the rearing of replacement queens as in the case in which a queen is lost. This time, their preparation for producing new queens is not undertaken in panic. During a process of supersedure, the queen will be replaced by one of her daughters after the daughter has successfully completed a mating flight and initiated oviposition.

5.9. Disinformation and deception

All systems of communication are susceptible to being subverted by miscommunication and deception. This is also the case for the seemingly neatly ordered set of interactions that we have been describing for queen honey bees. These forms of subversion and deception emanate from false queens and other misfit workers that activate their ovaries even in the presence of a fully functioning queen (see Chapter 6). For all these cheaters among the workers, how do they circumvent the message from the queen and the brood that so efficiently controls worker fecundity? Workers respond to the queen's signal by exhibiting the range of worker behaviors that produce a well-ordered colony. In addition, this signal causes the ovaries to remain in an inactive state. All of these effects are achieved by a level of queen pheromone that is above the worker suppression threshold. However, this threshold is variable among workers and also varies with the experience of individual workers. Thus, workers that emerge into a colony without a queen experience a level of queen pheromone that is far below their threshold. Hence, they can start activating their ovaries, altering their behavioral repertoire, and raising their response threshold toward queen and brood pheromones. This threshold may ultimately be raised to the point where they are no longer responsive to any inhibiting signal and have escaped from the queen's manipulated control. At this point, they become laying workers and false queens. In queenright colonies, there are two nonexclusive possibilities for individual workers to become false queens. First, the physiological threshold for pheromonal suppression is elevated. For example, the perception of pheromones might be impaired in such workers due to genetic variance. They simply might be chemically deaf to the pheromones of the queen and brood because they lack the appropriate receptors. Second, workers can avoid the vicinity of the queen and the brood nest. Indeed, reproductive workers have been shown to prefer to be away from the queen and never join the retinue. In response to a low exposure to pheromones, they might be able to produce their own pheromone signal, thereby increasing their response threshold to the point where they are insensitive to suppression by others.

In the case of the queens, probably as a result of their more rapid development during the pre-adult stage, they do not have any suppression switch and hence are not affected by their own mandibular signal. This may provide an explanation for the puzzling question of why queens are not suppressed by their own signals. The entire relationship between the queen and her workers is regulated by this dynamic system of feedback between the signal produced by the queen and the variable response thresholds of the worker population to the signal. Whereas most workers will be suppressed by the queen pheromones, there will be a small number that can evade the effects of these pheromones.

{6}

The Worker Bee in a Variety of Guises

FIGURE 6.1 *An alert worker bee.*

Ey! Goddes mercy!" sayd our Hoste tho,
Now such a wyf I pray God keep me fro.
Lo, suche sleightes and subtilitees
In wommen be; for ay as busy as bees
Be thay us seely men for to desceyve,
And from a soth ever a lie thay weyve.
And by this Marchaundes tale it proveth wel.

—Geoffrey Chaucer, *Canterbury Tales* (the Squire's Tale; 1386–1400)

6.1. The busy bee

In Geoffrey Chaucer's days, it was very clear that bees were busy but also that they were not necessarily efficient (Figure 6.1). It took a few more centuries to gain a quantitative understanding of the honey bees' efficiency. Indeed, the evidence seems to prove good old Geoffrey wrong. The impressive temporal division of labor has been dissected with all possible biological tools from the phenotype down to the individual gene cascades in the genome. The beautiful interactions among workers have been the focus of honey bee research over centuries. A clear highlight was awarding the Nobel Prize in Physiology or Medicine to Karl von Frisch for unraveling the mysteries of the dance language (von Frisch 1965). In numerous monographs on honey bee biology, the efficiency of workers had been a source of wonder, showing how they negotiate in a most adaptive way with their environment (Lindauer 1961; Seeley 1985, 1996), how they optimize house hunting during swarming (Seeley 2010), or how worker specialization is achieved through most efficient but yet inevitable self-organization processes (Page 2013). The mystery is clearly how a small insect with a miniature brain manages all these complicated tasks that allow the colony to work in an adaptive manner as a "superorganism" without some kind of central control (Moritz and Southwick 1991).

Indeed, it is subtle behavioral thresholds that govern when a honey bee worker engages in a certain task. This causes workers to specialize in specific tasks and not just engage in any random activity. A thought experiment may illustrate how coordination among workers and task specialization works without any central information system. Imagine a group of workers that are guarding the hive entrance (Figure 6.2). These workers are important because

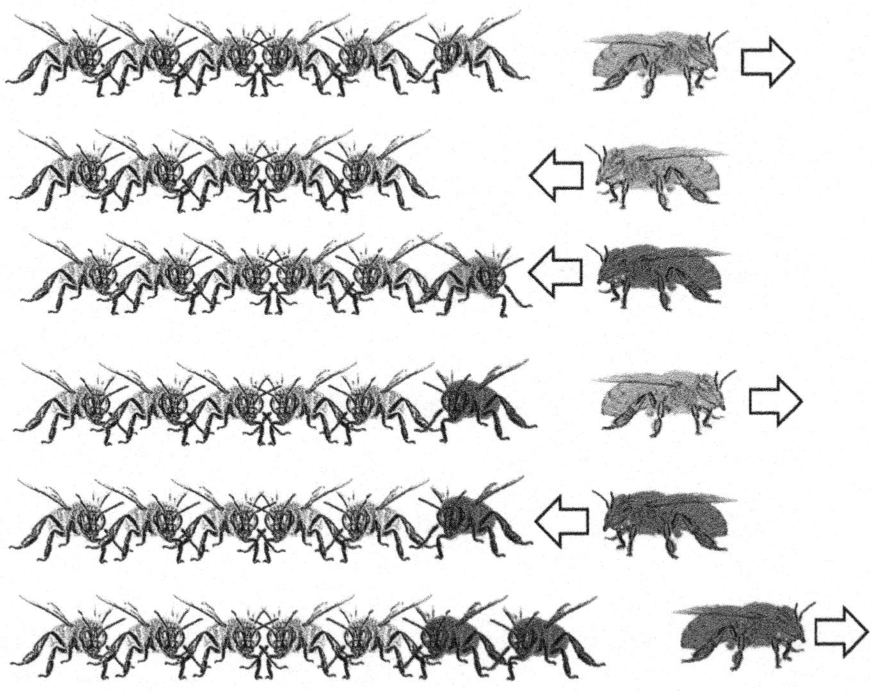

FIGURE 6.2 *A model for the self-regulation of the number of guard bees. Workers of the blue (dark grey) subfamily have a lower threshold for joining the guards. Eventually, most guarding workers will be of this subfamily. The phases from top to bottom are as follows: Six guards are sufficient for the red (light grey) bee, and she does not join the guards. If, however, one guard leaves, she will join the guard force because she has a threshold of five bees to join. The blue bee will also join because she has a threshold of six bees. In the end, only bees that have a threshold of six bees will form the guard force of the colony, whereas the others undertake other tasks (for a color version see tuck-in page between 142–143).*

they scrutinize incoming honey bees, determining whether they belong to the colony or not. They also screen for predators and alert the colony when there is a need to defend it. The number of guard bees is rather small, and it is a rare task in the colony. So why will an individual engage in guarding? Just imagine you are a bee passing by the flight entrance. You see six fellow bees sitting there checking incoming nestmates and scrutinizing the environment for misfits. You think, "OK; six security staff is just fine to keep me and the others safe." A few minutes later, you come by and find only five security staff on duty. You believe this to be rather risky and do not feel at ease. So you join the security team. A few minutes later, another bee passes by who also joins because she believes seven guards are needed to be safe. You think, "OK, that is fine with me," and you leave the security team because now there are too many and you go back into the hive, probably in search of some other tasks or just to hang out with others in some corner of the honey stores. Through these decisions by individual workers, the security group will be composed of those bees with a low threshold to join the guards. In the example, it will be those bees that join if there are fewer than seven guards. As a result, bees with this behavioral threshold (seven bees) will more often participate in the group of guards than will others. Hence, a very subtle difference in the threshold for engaging in a specific task will govern who is doing what in the colony. If the guards also have a fixed time commitment to guarding and then leave the task, this will result in a dynamic group of guard workers all having the low seven bee threshold for joining. But do not fool yourself! This is a virtual model colony explaining the principle but much different from a real hive entrance. In reality, there will be much more noise in the system driven by factors such as nectar availability, intruder intensity (Downs and Ratnieks 1999), and certainly stochasticity. Nevertheless, clearly the variance in behavioral thresholds among individual workers is a most potent driver of worker specialization in the hive, as most comprehensively illustrated by Robert E. Page in the *Spirit of the Hive* (2013).

It now seems clear how the colony works. The workers have a propensity to engage in a task whenever an opportunity is encountered to work for the good of the colony—a model that has been called the "foraging for work" theory (Tofts and Franks 1992; Tofts 1993; Sendova-Franks and Franks 1993). Workers are supposed to browse through the colony to engage in work whenever they encounter a feasible task they are fit to do or particularly good at doing. So they start a task as soon as there is an occasion to do so. In such a foraging for work setting, they actively seek and quickly find their places in the complex structure of the colony. They deal with issues when necessary, they tackle problems when they arise—they are just efficient at what they do, and they always do it whenever there is the chance to do so. Alas, the foraging for work theory received some critical commentary for honey bees soon after it was defined (Robinson, Page, and Huang 1994). And be serious—after having learned about this marvelous motivation to engage in work, and the beautiful concept of honey bee workers selecting those tasks that they are particularly good at performing, we hope that you might at least wonder whether this may be too good to be true for a honey bee. Think of it: An individual works whenever possible, is always motivated, probably adopting optimal foraging theory to seek the task for which it is optimally suited. You are right: As a human being with familiarity with social systems, you have every reason to suspect this might be overdoing it. What about a break? What about motivation? What about a vacation? We need sleep, don't we? Even robots eventually need to be recharged or serviced to perform a certain task. More important, what if there are bees with extremely high thresholds for almost all tasks and hence never engage in anything? Is this impossible in the hive social system? If you happen to be one of those readers who had to share an apartment with several flat mates, you will be familiar with such individuals in human societies with extremely high thresholds for any work. Who deals with the pile of dishes? Who empties the garbage bin? There are behavioral thresholds that can vary tremendously among different individuals. If this also applies to honey bees, would this be maladaptive? Would a colony fail because such high threshold "workers" that would not work much? If they existed, should the colony do something about them and discipline such lazy workers and set them to work? What can be done about cheating workers? What about social parasites that bluntly abuse the system?

Indeed, all of these behavioral permutations exist, as we detail later. So how does the colony get by, despite the burden of lazy cheaters? How many lazy bones could a colony tolerate to remain functional and efficient? Is the colony governed by a "law and order" system in which participation is enforced? Can colony organization possibly be self-organized if there are bees with very different propensities ranging from busy to lazy, from altruistic to selfish? How efficient can the colony be in light of such variability in worker efficiency?

Next, we address the aspects of worker efficiency by analyzing some tasks carefully to determine if the biological solutions and adaptations are, if not optimal, at least such that the colony gets by.

6.2. Sloppy builders

The combs with their hexagonal cells have long been considered as an optimal structure to obtain the most storage space out of the building material. Indeed, if all cells were prefect identical hexagons, the colony would have an ideal storage space. The issue of comb-building precision and space optimization has fostered a large field of research that has been comprehensively reviewed by Randall Hepburn (1986) and later again together with Christian Pirk, and Duangphakdee (2014). They come to the rather devastating conclusion that

> the shape of a honey bee cell does not have its celebrated regularity; its economy is a teleological myth. The entire history of the honey bee cell in natural history, geometry and philosophy is the story of centuries-old misconceptions. (p. 237)

So here we go again: The honey bees are apparently bluntly failing when it comes to proper building. Yet if one looks into a beekeeper's hive, the situation actually does not look that bad and the cells are rather regular across the combs. But here beekeepers assist by substantially aiding the bees in constructing a well-ordered nest. They provide straight parallel frames holding "comb foundation" consisting of sheets of wax where the cell pattern is embossed in a perfect fashion. So this is clearly a manipulated result. In nature, honey bees are on their own and have neither frames nor an embossed comb foundation. Indeed, as we know that nobody is perfect, why would we expect honey bee workers to achieve perfection in this rather demanding geometrical task. Those who remember their old school days in geometry will be familiar with the pitfalls of drawing perfectly shaped symmetrical hexagons. So how to get the bees out of this pickle? Carefully scrutinizing the cell patterns on a naturally built comb, not built on a beekeeper's frame, unfortunately reveals that no single cell is identical to another (Darchen 1956). It is the regularity of the overall cell pattern to the human eye, combined with a buoyant expectation of perfection in nature, that keeps the myth alive. Nevertheless, if we look at a freshly constructed comb from a distance, all cells seem to have a hexagonal shape. However, if we look more closely, things are different. The suite of irregularities among the individual cells reflects remarkable imprecision (or creativity?) of the builders. Cells vary in size, shape, and pattern on the comb. Cells are mostly constructed in either a vertical or a horizontal pattern (Figure 6.3). In 13 naturally built combs screened by Christian Pirk (personal communication), the horizontal pattern was less frequently used (2) than the vertical pattern (11). Irrespective of this natural variance, the comb foundation prefabricated by beekeepers uses the vertical cell orientation. In addition, workers sometimes build a tilted pattern and, very rarely, a rather appealing but less efficient rosetta pattern.

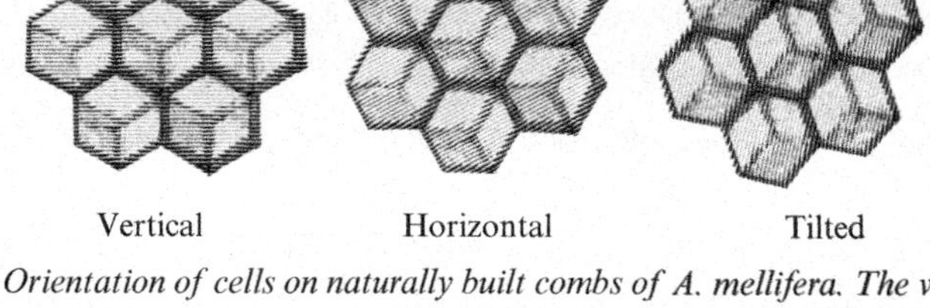

FIGURE 6.3 *Orientation of cells on naturally built combs of A. mellifera. The vertical and horizontal orientations occur at almost equal frequency, whereas the tilted orientation is very rare.*

In all cases, they seem to keep the hexagonal shape, which is why the hexagon is thought to be a hard-wired construction plan that all bees appear to follow. However, this hexagon is exactly why we were sweating in geometry class. How can the bees do this without ruler, compass, and protractor? The answer is rather demystifying: They do not make a hexagon in the first place. It is the physical properties of the wax and not the diligence of the workers that shapes the hexagonal cell. The bees themselves actually construct a rather imperfect round cell, sometimes even leaving spaces between adjacent cells. The workers do, however, elevate their body temperature while building the cells. The cell walls subsequently attach to each other, thereby transforming themselves to a hexagonal structure in response to its liquid equilibrium (Pirk et al. 2004; Karihaloo, Zhang, and Wang 2013) that is reached at close to 40°C (Buchwald, Breed, and Greenberg 2008). A major adaptive feature is not the outcome of meticulous construction abilities of the honey bee workers (they just build a crude circle) but, rather, the wax that does the trick by shaping the cell by itself into a reasonably symmetrical hexagon using the waxy cylinders provided by the bees.

This hypothesis has been challenged by others (Bauer and Bienenfeld 2013; Oldroyd and Pratt 2014); however, they, too, found that worker bees handling the wax had highly elevated body temperatures, with an average upper limit of 36°C. Although this is slightly below the liquid equilibrium of beeswax, one should keep in mind that these temperatures were recorded with a thermovision camera and hence represent the thoracic temperature of the back of the bee. The actual temperature of the wax in the comb under the bee may be slightly higher, allowing for the self-organized pattern formation. The work required to produce sufficient energy for deformation of beeswax decreases nearly sixfold with increasing temperature (230 Jm at 35°C to 40 Jm at 40°C; Hepburn 1986). Thus, even one or two degrees make a significant difference. We therefore believe that the self-organization hypothesis was supported rather than refuted by their results: The workers did enhance their thoracic temperature as predicted by the theory. Nazzi (2016) also found that the cells simply form themselves according to the number of cells surrounding them, which again gives additional support to self-organized cell formation on the comb rather than worker bees having a fixed template for making a hexagonal cell (Figures 6.4 and 6.5).

Indeed, all of this is not that new. Hepburn et al. (2014) remind us that the myth of honey bee cell precision was observed by Vogt (1911) more than a century ago, who concluded that

> the shape of the bee cell does not have its celebrated regularity; its economy is a teleological myth. The entire history of the bee cell in natural history in geometry and philosophy is the story of a 200-year-old mistake. (p. 67)

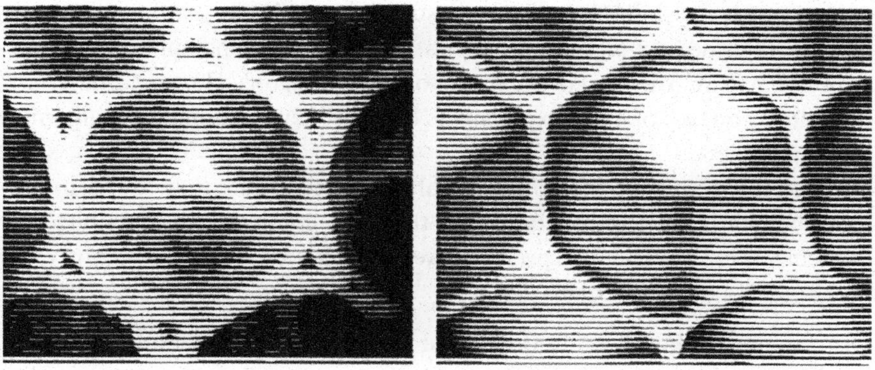

FIGURE 6.4 *Cell immediately after construction by the worker and after hexagonal transformation.*

FIGURE 6.5 *As soon as the regular pattern is lost on the comb, the cells' edges are determined by the number of adjacent cells. Here, a cell (asterisk) surrounded by seven instead of six cells takes the shape of an irregular heptagon. Also note the pentagonal cell just below, which is surrounded by only five cells.*

As we write, the mistake is now in its third century. There are still authors who praise the precision of the honey bee worker for cell construction, ignoring the biological evidence.

Darwin was also concerned about the perfection of comb construction by honey bees because the idea of perfection favors the argument from design. He conducted experiments and concluded the following (Darwin 1859):

> Thus, as I believe, the most wonderful of all known instincts, that of the hive-bee, can be explained by natural selection having taken advantage of numerous, successive, slight modifications of simpler instincts; natural selection having by slow degrees, more and more perfectly, led the bees to sweep equal spheres at a given distance from each other in a double layer, and to build up and excavate the wax along the planes of intersection. (p. 258)

His experiments showed that bees constructed round cells that became hexagonal with close packing, but he did not understand the potential of the ductility of wax in a warm environment to form hexagons. He would have been delighted to learn of these experiments because they undermine the argument from design.

Perhaps it is some sort of mystical sacred geometry that found its way into human consciousness that keeps the myth going. We keep finding such statements even in scientific peer-reviewed publications of the 21st century. For example, Kaatz (2006) states that

> bees are working with quite modest equipment, jaws, feet and sensory organs, and the fact that they can produce such well-ordered structures is amazing. The honeycomb is a macroscopic structure and is well-ordered in comparison with man-made arrays. In contrast, man uses all the technological advances possible, and produces some ordered structures as well. In the comparison of man with nature, the bees are not far behind.

The author, a physicist, was referring to ordered arrays in nanostructure science, which seems to weaken this argument. Given the honey bee comb may be a macroscopic structure at the human scale, it is certainly one at the scale of the honey bee. At our scale, the honey bee cell would hold the full body of a human spanning about 1 meter in diameter. Regardless of this scaling issue, clearly any sheet of beekeeper's wax foundation outcompetes the honey bee precision by far, and we tend to disagree with the statement, "In the comparison of man with nature, the bees are not far behind." In this context, it is actually a rather misleading statement in the first place because man is as much a part of nature as the honey bee. If the cells of the honey bee are substantially self-organized, a theory we currently favor, it is not surprising at all to find similar shapes of the cell on a honey bee comb and one in a nanostructured substrate that is also driven by self-organized mechanisms (Figure 6.6). Although the honey bees are less diligent than many readers might have hoped, the trick of taking advantage of the physical properties of wax is actually quite a good one that has misguided students of honey bee biology for thousands of years, as they searched for the way in which honey bees could make a hexagon and confound the geometry teachers.

6.3. The heroic bee

If honey is what honey bees are most famous for, then they are arguably infamous for their colony defense. Any predators are attacked with remarkable ferocity not just by single workers but in a most efficient mass reaction by thousands of defending workers. Indeed, the colony is in great need of efficient

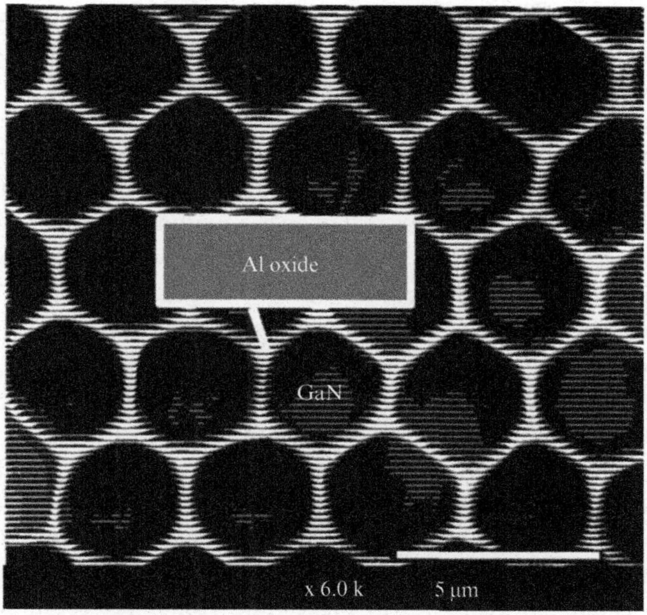

FIGURE 6.6 *Al-oxide honeycomb nanostructure on the n-GaN surface—a self-organized "honeycomb" nanogrid formed without any external agents.*

defense because it has precious stores of honey to which bears, but also greedy hominids (sometimes in the form of beekeepers), are absolutely addicted. Sugar is a high-energy food, and virtually all animals can use it as one of the highest quality diets provided in the wild. It is sterile, rich in carbohydrate, sweet, and golden—What more is there to say?

The price a predator or honey thief pays for a sweet honey treat is high, sometimes even lethally high. Honey bees, like all aculeate insects, have a sting that can inject poison into the aggressor. The sting of the honey bee worker is marvelously adapted to fight mammals that have a soft and elastic skin. The sting has a series of barbs that prevent it from being easily extracted once inserted into the aggressor. Conversely, stinging insect predators does not cause loss of the sting apparatus because the barbs do not hinder the retraction from the rigid cuticle. Also, when defending against arthropod attackers, the workers need to administer only very little poison for it to be lethal. Vertebrate predators receive the maximum poison load because the poison sack is also tightly attached to the sting apparatus and keeps pumping poison into the attacker, approximately 0.2 mg per bee (Schumacher, Tveten, and Egen 1994). Most of it is harmless water (88%) and also glucose, fructose, and phospholipids, which are similar to the hemolymph and pose no problem. The culprits are the other pharmacologically active components that have been described in great detail, including various enzymes, peptides, and amines. We shall not embark on a comprehensive discussion of the various effects of the individual components;

these are comprehensively analyzed in the proteomic study of Peiren et al. (2008). We mention only the main effects: It hurts because of histamine release by the victim, and it is primarily dangerous because of mellitin, the main peptide component.

For those who react with an allergic shock, a single sting is enough and the aggressor dies within one hour if not treated. But also for the non-allergic aggressors, there is a lethal dose of honey bee poison. The median lethal dose (LD_{50}) for an adult human is 2.8 mg of venom per kilogram of body weight. The authors of this monograph, who are fighting hard (alas futilely) to keep their 75-kg body weight, have a 50% chance of surviving 200 mg of bee venom (Schumacher, Schmidt, and Egen 1989; Schumacher et al. 1990). Hence, as little as 400 mg would be sufficient to kill both authors. However, this requires a total of at least 2,000 honey bees to kill one author on average, if both authors show up at the same time, if all stings are not removed after the attack, and if all venom sacks are completely emptied. Fortunately, even the seasoned beekeeper only rarely meets so many defending bees. So is it a lot? It obviously depends on the size of the defending army, particularly so because the honey bee army is one that relies heavily on recruits. Honey bee colonies can easily contain 50,000 worker bees. So an attack that killed both authors would cost approximately 4% of the colony population and is a cheap deal since the 2,000 workers required for this sad outcome could be replaced in one day by a fully active queen. Indeed, attacks of the Africanized honey bees that invaded the Americas have repeatedly reached such lethal levels, which is why they have been called "killer" bees.

But let us dissect a sequence of defensive behavior in more detail to understand the numbers of workers involved. The guards, a group of approximately 10–20 bees, are sitting peacefully although vigilantly at the flight entrance—until they recognize an intruder approaching the colony. This is done using optical cues. The guards immediately release alarm pheromone by exposing the sting and fanning to spread the chemical message. Next, they embark on an attack flight. Not all workers may sting, depending on the temperament of the individual bee. Indeed, in most cases, it is sufficient to ward off an intruder by simply circling the culprit with high-pitch, high-speed flights. Quite often, however, a few bees prefer to sting, and the intruder is now chemically labeled with a source of alarm pheromone and probably frantically trying to get rid of the nasty bees with swift movements, thereby spreading the volatile pheromone. The alarm pheromone will quickly recruit more workers from the colony in search of an intruder. Depending on their temperament, they will attack not only the intruder but also any moving object within the vicinity of the colony. This is neither particularly fair for any peaceful observer nor necessarily adaptive because all efforts should of course be concentrated on attacking the aggressor. The duration of the defensive response is also highly variable and again corresponds with the temperament of the honey bees. It can sometimes take hours before the colony calms down in response to the slowly decreasing levels of alarm pheromone. For the less experienced, the alarm pheromone, isoamylacetate, smells pleasant—a bit like bananas. However, mammal learning through conditioning by punishment is swift and long-lasting. The beekeeper's apprentice quickly learns to associate the smell of the alarm pheromone with the subsequent pain of a honey bee sting. Thus, if any reader happens to end up in such an unfortunate situation, it often pays to leave the place immediately. Sometimes even speed may matter and running might help. By doing so, you move the source of the recruitment pheromone (yourself with all the stings in your skin) away from the colony, which considerably limits the number of new recruits. Staying and making hectic movements trying to get rid of attacking bees is rather counteradaptive.

What seems an efficient defensive strategy at the colony level is very different when we view it from the level of the individual worker. The rapidly escalating colonial defense is based on a self-organized positive feedback loop. Workers sting, leave their sting to mark the attacker, which recruits more attackers—a clever strategy, or is it not that clever after all? The entire system appears to work because the workers kill themselves for "the sake of the colony." Because of the barbs at the tip of the sting, it will stick in the skin of a vertebrate victim. The sting apparatus autonomously inserts itself even deeper into the skin by the alternate movements of the lancets (Figure 6.7). Eventually, they are so deep in the skin that the sting cannot be retracted. Because the poison sack sticks to the sting apparatus when it is shed by a worker, the victim is now provided not only with the alarm pheromone that keeps being released from

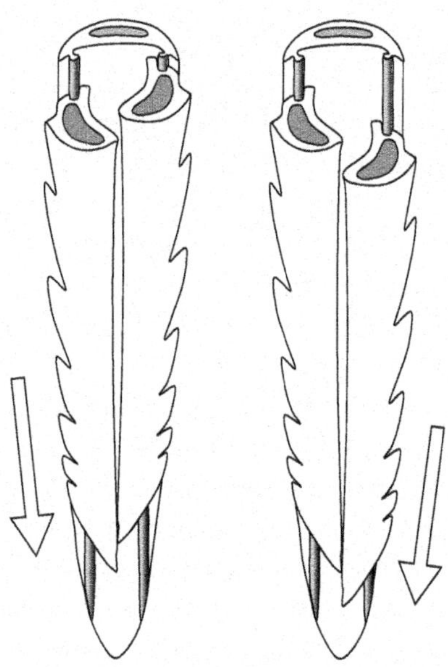

FIGURE 6.7 *Lancets entering skin, pushing sequentially deeper and deeper.*

the surface of the sting sheaths but also with the maximum possible dose of poison. The shedding of the sting causes the death of the worker because together with the sting apparatus, it loses its last pair of ganglia that innervate the sting apparatus. This causes the most unfortunate demise of the gallant defender. The classical adaptive explanation for this shedding is that the attacker receives the full poison load and is labeled with the alarm pheromone, and subsequent bees can selectively attack the intruder. However, the poison sack is almost completely emptied after 16 seconds (Schumacher et al. 1994), so to leave it behind is probably not very efficient after all, and multiple stings might be equally if not more effective in terms of administering poison. And why not cling to the attacker, thereby also releasing alarm pheromone, if one is going to die anyway?

Shedding of appendages does occur in a number of members of the animal kingdom. Animals shed their tails or spiders their legs to confuse predators and to avoid death. Why is heroic death only observed in the social insects (Hermann 1971)? Is it because workers are more expendable since there are so many of them? Indeed, for a solitary bee, it would not be adaptive to kill herself before completing the production of offspring. But then why do large colony social insect species not all exhibit this extreme altruism if it is such a tremendous evolutionary invention? Are you a reader who has been attacked by a yellow jacket wasp? Equally nasty, equally efficient, but wasps do not lose their sting. They have no barbed lancets. Do wasps still exists? Seems so—and some are most efficient invasive species, such as European *Vespula* species throughout the world (Beggs et al. 2011). So perhaps the barbed sting is not such a clever and adaptive solution by the honey bees after all? Even workers of the Asian honey bee *Apis cerana* have less well-developed lancet barbs on their stings, allowing them to occasionally withdraw and sting again (Sakagami and Akahira 1960).

Self-sacrifice can be a tactic in warfare, but it is arguably not a very clever one and therefore often described as brave. Many human societies have repeatedly been able to convince their male youth that lethal self-sacrifice is good for society. Alas, history has shown that this tactic has rarely, if ever, been decisive at a strategic level. In animal systems, it is only successful if the sheer number of attackers vastly outweighs the opponent. The logic of why success is so difficult achieve seems self-evident: You can strike only once, but this may not be lethal to the opponent.

Yet all honey bees share the trait of the barbed sting in workers, and given that the genome provides for a phenotype with a sting without barbs in all species (the queen), it should have been easy to get rid of this trait if it had been not adaptive. Indeed, it may be the specific predation by mammals that makes it work. Only a few species predate on wasp nests, which is not surprising because they do not taste particularly appealing, lacking the sweet honey. So sting autotomy may be an adaptation to protect the honey store. The barbs are not adaptive for the queen, which uses its sting primarily to kill rival queens but can also sting vertebrate intruders. It is clear that the cost to the colony of losing the queen would be too high just because of a few barbs on the sting. The possibility for sting retraction is therefore essential for the queen caste, but the adaptive value for workers may remain less clear than one might have hoped.

6.4. The policing bee

Although the honey bee society has been portrayed repeatedly as one of utter harmony and efficiency (Seeley 2010), we have already expressed our doubts about so much goodness among biological organisms. Evolutionary biologists were delighted when they understood why all workers working for the good of the colony is an evolutionary stable strategy through the fundamental work of William "Bill" Hamilton (1972 Box 6.1), the first and most important advocate of inclusive fitness theory. The close relatedness among the females in haplodiploid social insects sets the stage for the evolution of altruistic behavior. Although his theory brilliantly explained why worker sterility in male haploids is an evolutionarily stable strategy, it quickly became clear that the honey bee was a lousy and completely unsuitable model system to provide useful empirical support for inclusive fitness theory. The extreme multiple mating (polyandry) of the queen and the mixing of semen in the spermatheca result in exceptionally low relatedness among the workers within the colony. In fact, the honey bee holds the world record for low intracolonial relatedness in a monogynous colony due to the excessive number of matings of the queen, which can be in the range of 100 drones per queen. Whenever a worker meets another one in the hive, she will in most cases encounter a half-sister rather than a super-sister. This has many consequences for what is adaptive for a worker to do and what not to do.

BOX 6.1 Inclusive Fitness Theory

The existence of sterile individuals has presented evolutionary theorists with a conundrum regarding how these individuals could evolve since they produce no offspring. Without offspring, they should immediately die out. For Charles Darwin, (1859) this was a matter that required particular attention when developing his evolutionary theory based on natural selection, which was predicated on those with a higher reproductive fitness prevailing over those producing fewer offspring:

> [I] will confine myself to one special difficulty, which at first appeared insuperable, and actually fatal to the whole theory. I allude to the neuters or sterile females in insect-communities; for these neuters often differ widely in instinct and in structure from both the males and fertile females, and yet, from being sterile, they cannot propagate their kind. (p. 236)

Darwin got around this problem by explicitly invoking family selection using relatedness as a major argument:

> Hence I can see no real difficulty in any character becoming correlated with the sterile condition of certain members of insect communities: The difficulty lies in understanding how such correlated modifications of structure could have slowly accumulated by natural selection.
>
> This difficulty, though appearing insuperable, is lessened, or, as I believe, disappears, when it is remembered that selection may be applied to the family, as well as to the individual, and may thus gain the desired end. (p. 237)

Yet when it became clear that heritable genes determined the phenotypes on which natural selection was operating, a genetic model needed to be derived that could plausibly explain the evolution of traits reducing reproductive fitness of some in favor of others. It was William Hamilton (1964) who first realized that relatedness as an inevitable property of any biotic system is fundamentally important for a genetic evolutionary theory. Because biotic systems produce offspring that share genes among themselves and with their parental organisms, sharing genes (i.e., relatedness) is an inevitable biotic property and is the very basic foundation of any form of life. So if an individual supports and enhances the reproductive fitness of a relative, this will be evolutionary successful whenever the costs (c = reduction of fitness of altruist) multiplied by the relatedness to the recipient (r) are less than the benefit (b = enhanced fitness of recipient). Thus, whenever $b > r \times c$, altruistic behavior should be adaptive and favored by natural selection. The high frequency of the independent evolution of sterile worker castes and eusociality in insects strongly supports this theory. The vast majority of eusocial species belong to the order Hymenoptera, whose species are characterized by male haploidy. So if a diploid female is mated with a single haploid male, all her female offspring will be related by 0.75 (super-sisters). The benefits resulting from any altruistic support of a super-sister can be much smaller than in diploid species, in which sisters are related by 0.5.

In highly derived eusocial Hymenoptera, relatedness among the workers of a colony can be very small because colonies can have many unrelated queens (polygyny) or queens can mate with many males (polyandry), as is the case in all honey bee species. This does not reject Hamilton's theory because once insect societies form long-lived colonies, selection at the colony level becomes increasingly important in comparison to individual worker fitness. The evolutionary-relevant phenotype shifts from the individual to the colony, a principle clearly foreseen by Darwin in *Origin of Species* (1859).

Let us examine the case of egg laying, which seems to be extremely attractive for the worker because she is of course much more closely related to her own sons than those produced by the mother queen. Alas, she is not the sole worker in the colony. Her sister workers might not be impressed by finding worker-laid eggs in a cell that were not laid by themselves. In fact, because there are so many subfamilies in the colony, the chance of encountering an egg laid by a

FIGURE 6.8 *Policing worker with a worker-laid drone egg removed from a cell.*

half-sister rather than a full sister is extremely high. So unless she can discriminate kin (super-sister-laid egg) from non-kin (half-sister-laid egg), it would be at least highly adaptive to discriminate between a worker-laid egg (average relatedness, ~0.25) and mother queen-laid male eggs (average relatedness, ~0.5). Thus, as soon as workers cannot reproduce themselves, they should prevent other workers from reproducing and certainly kill any offspring produced by laying workers, a process termed "policing" in social insect biology. This occurs in honey bee colonies (Figure 6.8). Fewer than 1% of all worker-laid eggs eventually develop into mature drones. The vast majority are removed and killed by sterile workers, and this has been presented as evidence for effective worker policing (Ratnieks and Visscher 1989).

How "clever" (i.e., adaptive) is this policing strategy really? From inclusive fitness theory, we know that supporting your kin may be rewarding if the benefit (b) exceeds the costs (c) in relation to the relatedness (r): $b > c/r$. We also know that intracolonial kin recognition is in principle possible in social insects, and the chemical composition of the hydrocarbons on the cuticle varies among the various subfamilies in a colony (Nehring et al. 2011; Helantera and d'Ettorre 2015). For example, Stuart and Herbers (2000) report on kin recognition by assessing patterns of aggression within and between nests of the ant *Leptothorax longispinosus*. So why don't honey bees do this? Because if they could, a worker should not bother to become a police woman in the first place. The worker should discriminate between an egg laid by a super-sister and an egg produced by a half-sister: Care for the first and kill the latter. In fact, in order to maximize its individual inclusive fitness, it should defend a super-sister egg against potential policing from other subfamilies and prefer to nurse its own patriline's worker-laid male offspring (r = 0.75) rather than the queen-produced males (r = 0.5). The patrilines should start establishing territories inside the nest that accept the queen but not workers of other patrilines. There should be territorial conflict over brood cells, with workers watching jealously over their subfamilies' cells. Overt fights would be the result of such a development. Of course, this scenario would have a severe impact on colony efficiency for all its members and hence reduce the indirect fitness gained from the colony. In principle, the lack of conflict can therefore easily be explained within the framework of indirect fitness gained from the colony (Ratnieks 1991; Boomsma and d'Etorre 2013). Only if conflict is kept in check will there be reproductive output in terms of colony growth that eventually can turn into reproductive swarming and the production of sexuals. But how did evolution solve this problem between individual, family, and the colony interest? It obviously selected for workers that were unable to discriminate kin at the brood level. Evolution selected the less smart ones, those with limited sensory capacity, because clearly larvae of different subfamilies can in principle be easily recognized using the heritable differences in cuticular hydrocarbon composition. Many social insects have been shown to use these cues, and certainly gas chromatographs can produce subfamily-specific peak profiles in honey bees (Ogden, Pardo, and Tchalpa 1998; Arnold, Quenet, and Masson 2000). Alas, this rich source of information is obviously not used by the honey bee workers. There are no reports of clan fights in the colony. Despite a suite of critical experiments, it seems the honey bees do not use their sensory system to discriminate kin inside the colony (Moritz 1991; Kryger and Moritz 1997; Rangel, Mattila, and Seeley 2009). However, there are very detailed accounts of laying workers being harassed or even killed inside the colony (Anderson 1963; Dampney, Barron, and Oldroyd 2002). There are also excellent studies quantifying egg removal by policing workers (Ratnieks and Visscher 1989). So whether we like it or not, the police took over in the honey bee colony—not necessarily the most sensitive bees though (we explicitly do not compare the bees with human systems).

6.5. The lazy bee

Students of honey bee behavior have been impressed for millennia by what honey bees can do. Far fewer of them have considered what honey bee workers do not do, although they might be able to do something useful. Studies that examined workers engaging in no tasks of course found the inevitable: The frequency of workers doing nothing in the colony exceeded the frequency of those engaged in a range of tasks by far. Indeed, Lindauer (1961), who followed a single honey bee throughout its life, observed that this bee "loafed" 40% of the time rather than being busy. Years later, Seeley (1995) confirmed this observation and reported a high rate of inactivity especially in non-forager bees, resulting in approximately one-third of all workers in the colony resting at any given moment during daytime. Thus, doing nothing seems to be a very frequent state for honey bees. Anderson (2001) examined the issue of the non-active worker by using a very simplified model that discriminated between a successful and an unsuccessful scout as well as a recruited and an inactive worker. Despite the simplicity of the model, he could show that the inactive workers serve as an important reservoir to cope with situations in which nectar flows are patchy with regard to time, space, and nectar quality. However, how many of these reserve bees does the colony need? Understanding that reserve bees are convenient when needed is one thing; testing whether the investment in these reserves is worthwhile and cost-efficient is another. Indeed, van Nest and Moore (2012) report that much higher numbers of active bees are required to achieve optimal forager allocations than observed in reality. However, all current models are deficient in that they do not include the colony size, which is clearly a major factor driving the realization of an optimal solution for worker allocation. Moreover, the number of combs and amount of available nesting space are known to play a central role in worker allocation. Foraging is just one aspect of the equation because incoming nectar and pollen need to be stored in empty cells, and space must be available to expand storage capacity in the nesting cavity (Camazine and Sneyd 1991; Camazine 1993). Thus, here is a research field begging to be addressed that deals with the costs and benefits of a reserve worker force. It is probably prudent at this stage to leave this question unexplored and simply appreciate the empirical evidence: Bees remain less busy than many who subscribe to the belief that the honey bee colony is a "perfect" society, in which every member works in the common interest of all the members of society, would have hoped—a philosophy identified as an illusion many years ago by Karl Marx (1845).

Honey bees are primarily active during the day, and they have distinct circadian rhythms, with phases of activity and rest like most diurnal organisms. Although there are occasional reports of African and Asian honey bees foraging at night using moonlight for orientation (Fletcher 1978; Warrant, Porombka, and Kirchner 1996), the careful observer will often see workers hiding in a cell in which they exhibit all the characteristics of human sleep (Figure 6.9). Whereas humans perform rather regular breathing during sleep (and some even refer to the authors as snorers), sleeping bees show discontinuous ventilation of their bodies. Long periods of immobility are interrupted by pulses of dorsoventral contractions of the abdomen, which is when gases are exchanged (Kleinhenz et al. 2003; Sauer et al. 2003). Nevertheless, although honey bees have not been reported to snore, their brain activity

FIGURE 6.9 *Phases of sleep outside the cell: left-awake bee with raised antenna. These are lowered as the bee gets more sleepy to eventually touch the comb on the far right. The bee in deep sleep also lowers its body posture such that the abdomen touches the substrate.*

shows the unique patterns of sleep we know from higher vertebrates (Kaiser 1988). In this sleeping state, bees also show a reduced responsiveness to disturbances.

However, bees do not only sleep at night. Indeed, Karl von Frisch (1940) complained about the sleepy society when the workers did no foraging despite perfect weather conditions. He confirmed the reports of von Buttel-Reepen (1900) that workers adjusted the timing of their daytime activity in relation to the time of available forage. If there are phases of activity, there are also phases of inactivity. So do they take naps during the day? It actually seems foragers try to sleep whenever they can (Klein and Seeley 2011), and this might strike a sympathetic cord with many readers. This seems to be highly adaptive because workers with sleep deprivation suffer substantially. They have impaired communication (Sauer et al. 2004), degraded memory formation (Hussaini et al. 2009; Klein et al. 2010), and seriously reduced orientation skills (Beyaert, Greggers, and Menzel 2012)—very similar to higher vertebrates, including humans.

Sleep duration is also dependent on the task specialization of the workers. For example, cell cleaners appear to be the most sleepy members of the hive (Klein et al. 2008). It is these young bees that prefer to sleep inside the cells with long unbroken phases of sleep. This very obvious sleeping behavior becomes less frequent as the workers age. The old foragers sleep outside the cells and preferably during nighttime. They are the ones that express the most distinct diurnal circadian activity. Foragers have highly expressed circadian activity, which may be an essential requirement for them to express the precise timekeeping abilities necessary for foraging and dance communication (Bloch 2010). Foragers even adjust their foraging flights to show a food-anticipatory activity—that is, they fly out to the previously rewarding food source earlier than the day before so as not to miss the opening of the flowers and be the first ones to harvest the nectar. There are foragers that seem to know when to seek specific forage in space (Pahl et al. 2007). However, not all worker bees show the same behavior in relation to timekeeping (Wagner et al. 2013). Foragers even show activity preferences that depend on the patriline to which they belong and can be classified into early and late shift workers (Kraus, Gerecke, and Moritz 2011). This matches the observation that foragers take small bouts of sleep during the day by adjusting their daytime sleep in relation to their previous foraging experience (Klein and Seeley 2011).

In contrast, nurse bees have been shown to be active around the clock and seem to lack any circadian rhythms (Crailsheim, Hrassnigg, and Stabentheiner 1996; Moore et al. 1998; Toma et al. 2000; Bloch, Rubinstein, and Robinson 2004; Bloch, Shemesh, and Robinson 2006; Shemesh, Cohen, and Bloch 2007; Shemesh et al. 2010). However, as soon as they are removed from the hive (e.g., experimentally), they start expressing circadian rhythmicity. Whereas the adaptive value of a circadian activity in foragers seems obvious, it is less clear how efficient it is for a nurse bee to have no rhythmic activity. Although their activity is required during day and night, clearly one would expect them to also go through periods of sleep. Wouldn't it then be more efficient to work in shifts rather than random time slots for a task that is required throughout the day to maximize the availability of nurses for the brood? Nurses inspect brood cells to feed the larvae and also feed adult nestmates (Crailsheim 1992). So clearly a pattern of rest and activity is to be expected for the individual nurse bee. Rodirguez-Zas et al. (2012) report on various gene transcripts that show circadian abundance oscillations in nurses. Klein et al. (2014) observed that nurse bees sleep more than 10% of the time outside the cells and another 15% of the time in cells. So clearly there are phases of rest for individual workers, but these are not coordinated with the other nurse bees so that overall circadian patterns at the behavioral or neuronal level are not detected. By contrast, stingless bee nurses (*Melipona quadrifasciata*) did show circadian rhythms (Teixeira, Waterhouse, and Marques 2011), although with amplitudes an order of magnitude less than those of older workers. Nagari and Bloch (2012) suggest that constant pheromonal signals from the brood prevent nurse bees from expressing circadian activities because deantennated nurses immediately switch to circadian activity. Thus, although there is no common circadian pattern for all nurses, there is still room for the individual nurse bee to develop its own pattern of rest and activity that eventually might also develop into a shift-like pattern of work as observed for foragers (Kraus et al. 2011). It may be the sheer number of nurses available in relation to the brood in combination with a self-organized individually based threshold for feeding larvae by the nurse bee that may do the trick. A mechanism based on a constant larval signal for food and an individual depletion of glandular food jelly secretions that need to be replenished by the nurse bee during a phase of sleep might well be the driver the nurse bees need to develop a shift-like activity pattern. They would complement each other's availability, resulting in the larvae being fed 24 hours a day. The ratio between larvae and nurse bees would then determine the phases of activity and rest. Although this is a model of considerable appeal to both authors, we warn any reader: It clearly requires both a solid theoretical model to show this is functional in principle and, much more important, comprehensive empirical testing to show that what is possible in theory takes place in reality. Overall, however, there seems to be more sleep than needed to keep every bee in the hive alive and healthy.

6.6. The stupid bee

Honey bee workers have provided one of the most widely explored experimental model systems for understanding principles of learning and memory. They excel when it comes to learning odors and colors. They have exceptional capacities to learn and remember spatial and temporal patterns that allow them to navigate in the most complex environmental and experimental settings. Numerous volumes and publications deal in-depth with the behavioral and neuronal mechanisms that enable the bees to outperform many higher vertebrates with regard to odor and color learning. However, although the miniature brain of the honey bee manages amazing cognitive competence, clearly there is individual variance among bees. Not all bees are equally smart and they do not have similar individual experiences, which inevitably results in considerable individual variation with regard to communication, learning, and memory performance (Biesmeijer and Seeley 2005; Leadbeater and Chittka 2007; Wei and Dyer 2009). These individual differences in cognitive ability are discussed in the context of enhancing colony performance. For example, different foraging experiences of different workers are exploited to enhance colony efficiency (Dornhaus et al. 2006). Karl von Frisch (1965) was the first to report on the "Missweisung," small deviations expressed in the dance communication. Although there have been attempts to interpret the imprecision in the dance language as an adaptation to specific environmental conditions, Tanner and Visscher (2010) concluded that it was more likely precision constraints that drive the sloppy dance language. Seeley (1995) observed that "on average, a bee will need to make approximately four tries, that is, conduct some four dance-guided searches, to locate a flower patch advertised by a dancer" (p. 126). So there are bees that can do with fewer or more dances before they arrive at the right spot. When experimentally challenging the workers to conduct more complicated tasks under more controlled conditions, the individual differences become even more obvious. There was large variance in learning speed among honey bees that were tasked with performing reverse learning (Mota and Giurfa 2010). Dyer et al. (2014) used the reverse learning paradigm to illustrate the consequences of having workers with different learning behavior in the colony. They identified three different behavioral classes of worker bees:

1. "Deliberative-decisive" workers that were able to make a major change to the initially learned preferences
2. "Fickle-circumspect" bees that changed their preferences only by a small amount in every reversal learning act
3. "Stay" bees that refused to show reversal learning and did not change from their initially learned preferences

So, clearly we find the full range of behaviors, from an opportunistically reactive bee to a stubborn bee failing to be impressed by any change. Some readers, having been involved in teaching for many years, will find this categorization familiar, and it is certainly not an unknown degree of variability in mankind. However, how good is it for the colony to have some workers learning less well than others?

Dyer et al. (2014) used agent-based computer simulations including parameters describing flower reward switch oscillation frequency, flower handling time, and frequency of flowers with no reward. They found that Fickle-circumspect bees are more efficient at nectar collection under high frequencies of reward reversals. However, as reward reversal frequency decreased, the performance of Deliberative-decisive bees became more efficient. At low reversal frequencies, Fickle-circumspect and Deliberative-decisive bees became equally efficient. Remarkably, the stubborn Stay bees never outperformed the two other groups. Although the authors concluded a fitness benefit for honey bee colonies containing individuals exhibiting different strategies for managing changing resource conditions, the role of the stubborn ones remains rather elusive and a burden on colony efficiency.

But it is not only foragers that have been studied. Guards have also been shown to be less efficient than one might hope. Couvillon et al. (2013) reported that the error rate in nestmate recognition of individual guards at the flight entrance exceeded 30%. Although this may seem unacceptably high, it is actually not so bad because of the group of guards involved. There is more than just a single guard worker at the flight entrance, and the chance of detecting a foreign bee increases rapidly with increasing guard group size. With a five-worker guard group, the chance of an intruder passing by unrecognized is as little as 0.24% $[(0.3)^5]$ because one worker in the group may detect the intruder and then alert the others. This is an example of individual workers performing less than optimally, but this is not important at the colony level because there are many workers performing the same tasks. For many colony-level adaptations, the colony can afford to have non-optimal workers simply because of the sheer number of workers involved.

Given that the failure rate of an individual worker may be 100%, others will compensate and the large number of workers in the colony will ensure that the colony remains operational. So yes, there could be many very stupid individuals specializing in specific tasks. However, the high redundancy in the colony ensures success, which is typical of the functioning of complex systems in general.

6.7. The stealing and robbing bees

Stealing resources produced or gathered by others is not just common in biological systems, it is the very basis of any animal life because animals are heterotrophic. Hence, stealing and robbing (i.e., stealing by force) are also well known in honey bees.

6.7.1. STEALING FROM PLANTS

Plants provide food for honey bees. They do not do this because they love bees but, rather, because they require a pollination service in exchange. The plant's plan is to attract the honey bee by its sweet nectar and then contaminate the furry hair cover with its pollen. The pollen is then transferred to the stamen of the next conspecific flower. Sometimes the flowers make it difficult for the bees to reach the nectar to make them stay longer and work harder in the flowers to reach the nectar. This ensures that the pollen is really transferred.

However, not all bees comply with the rules set by the plant. Particularly when no mass nectar flow is available, honey bee workers can become most creative in extracting nectar from a flower while overcoming the pollination plan developed by the plant. Workers can steal the nectar without paying the pollination price. They do this by cutting a hole at the base of the flower, allowing for direct access to the nectar—a process called nectar robbing(Figure 6.10).

In addition, there are reports of rare observations that honey bee workers act as highway robbers and steal pollen from other bees that have collected pollen from flowers. However, this seems to be a rare exception (Laroca and Winston 1978; Thorp and Briggs 1980; Jean 2005). Occasionally, honey bee workers have been observed to steal the pollen from the corbicula of bumblebees.

6.7.2. ROBBING FROM COLONIES

Whereas stealing and robbing from individual bees seem to be the exception in honey bees, it is the socially organized robbing raids that matter at the colony level. Honey bee colonies provide resources carefully collected by their workforces, ready-made, to be used at any time by a resourceful robber. In particular, the honey stores are not just appealing to large mammal predators but are also most attractive to other honey bees. Indeed, honey robbing is probably more frequent between honey bee colonies than that undertaken by

FIGURE 6.10 *Worker stealing nectar without providing a pollination service.*

the odd bear or honey badger. The frequency of colony-level robbing under natural conditions is not known. It will certainly depend on the local density of colonies within the foraging range of the workers. The apiary setting arranged by the beekeeper therefore greatly facilitates robbing among colonies, and robber bees are well known to apiculturists. Robbing occurs during periods of low forage availability when honey bees become desperate in their search for food. Clearly, the hive next door may have some rewarding honey stores if their own stores are running low. So the old foragers start searching and will try to gain access to the neighboring colony. In this period of the season, it therefore pays to check incoming bees with particular attention. When a robber is successful in breaching the defenses of an adjacent colony, it will bring home its precious load and immediately communicate this to its fellow foragers. This produces consequences that beekeepers dread. Because the colonies in the apiary are all very close to each other, the dancing robber communicates her find with a round dance not indicating any particular direction. Hence, recruits start eagerly searching for honey sources in the immediate vicinity, not just the colony that had been spotted by the first successful robber. Whereas some of these recruits may end up in the same colony, other bees may now fly to another colony in the apiary, trying to pass the guards and get access. There will be heavy fighting between the guards and the robbers. As a consequence, all colonies in the apiary are at risk of robbing, and particularly weak colonies are at risk of being overrun. Robbing is a frequent cause of colony loss arising from beekeeping practice. Although this is clearly a very dark side of honey bee behavior for the beekeeper, under natural conditions this is a most adaptive behavior. It is classical Darwinian selection with all its brutality at the colony level: death of those that cannot defend themselves.

6.7.3. STEALING FROM BIRDS

A most extraordinary form of robbing was observed by us when we noticed a large number of honey bees hovering over a bird-feeder dish filled with bird seed. At first, we suspected the workers were hovering over the feeding dish trying to collect residual dust from the feeding trays. However, a closer inspection revealed that they were actually collecting whole seed kernels (Figure 6.11).

Rather than collecting the dust and compressing it in the pollen baskets, they simply rolled in the bird seed until eventually a grain would by chance be caught by the corbiculae. Once both corbiculae were loaded with a seed, they returned (we do not know if proudly) to the hive, where we were unfortunately unable to determine the fate of the seed grains because the colony was behind the foundation of a house, just underneath the bathroom. Given that the grain foragers appeared in large numbers at the bird feeders, it seems

FIGURE 6.11 *Workers stealing seed kernels from a bird feeder. The flying bee has both corbiculae loaded with the seed kernels.*

obvious that they must have actively recruited fellow workers from the hive to join in the foraging effort. This would require recruitment dances of the experienced workers from which the followers learn about the positions and the quality of the food source. However, it seems highly unlikely that the bees can actually use the seed grains as food. The seed is covered by the hard shell of the seed coat, likely leaving the ingredients of the kernel inaccessible to the bees. Because pollen foragers deposit their loads directly into the storage cells, they receive no feedback from workers in the hive on the quality of their forage. Pollen collection is therefore more readily prone to errors as in this exceptional case of the theft of bird seed than in the case of nectar foraging. Here, the workers that return from foraging flights pass the collected nectar over to in-hive receiver bees that provide an additional filter with regard to food quality.

6.8. The default reproductive worker

Whenever the queen dies, the colony is in need of a replacement. Normally this is regulated by emergency queen rearing. An egg initially destined to become a worker will be fed by the nurse bees with the royal jelly diet and a queen will be reared. Sometimes, however, this emergency rearing may fail, and the colony has no brood with which to requeen itself—a condition called hopelessly queenless. In these colonies, laying workers will arise after a few weeks and produce drone offspring because the eggs will not have been fertilized and are haploid. From a colony perspective, this looks adaptive because at least some sexuals will be produced before the colony eventually dies out. These laying workers not only activate their ovaries but also produce a queen-like pheromone signal that prevents ovary activation in other workers. Hence, only a few workers will develop into such "false-queens," as Sakagami (1958) termed these workers. At the individual level, this may seem to be an unfair process because only a few workers win the reproductive prize. However, if we were to envisage a situation in which all workers started laying eggs at the same time, this would clearly reduce the inclusive fitness of all individuals in the colony. Larvae need to emerge from the eggs laid and need to be reared to adult drones. This requires many nurses and workers to maintain the internal homeostasis of the colony. Hence, workers developing a pheromone signal the fastest will be those that win this intracolonial competition. This is not just by chance, and so-called anarchistic worker bees have been shown to have little respect for queen pheromones and sometimes even lay eggs in the presence of the queen.

A very different case of adaptive laying worker reproduction is found in the Cape honey bee, *Apis mellifera capensis*. This is a subspecies of honey bee that is endemic to the southern tip of Africa and specific to the fynbos region. Cape honey bee workers have a high potential to activate their ovaries and develop into laying workers. Another unique trait is that they produce female offspring although they are not mated. This type of parthenogenesis is termed *thelytoky* in contrast to *arrhenotoky*, the production of males. Arrhenotoky is the rule in all other subspecies of the honey bee, in which laying workers produce haploid drones. A specific aberration of the spindle apparatus during meiosis prevents the separation of the two central meiotic products in laying *A. m. capensis* workers. This re-establishes the identical diploid genotype of the mother in the offspring that has been called a pseudo-clone. There is no free recombination of chromosomes, and even crossing over is strongly suppressed. The sex locus always remains heterozygous, and these workers therefore produce females, more workers but also queens depending on the way in which their larvae are reared to adults. This results in a most efficient path to requeen a colony, even if the colony is hopelessly queenless. The laying worker of *A. m. capensis* can lay female eggs that are regularly reared to new queens. We showed how efficient this trait is in rescuing the colony and saving it from a takeover by a foreign queen clearly to the fitness benefit of all workers in the colony (Moritz et al. 2011).

Although this is a highly adaptive way to requeen a colony under hopelessly queenless conditions, the trait has also generated a preadaptation for social parasitism. Thelytokous reproduction in combination with a high potential for worker reproduction opens an additional life history path for Cape honey bees (Box 6.2). This social parasitic path is highly adaptive but a bane in the lives of apiculturists, as discussed next.

BOX 6.2 George Onions, Discoverer of Thelytoky in Cape Honey Bees

George William Onions, 1867–1941.

George Onions was born in Cape Town, South Africa, in 1867.[1] He was the son of an engineer who migrated to South Africa in 1859. Onions' mother and elder brother died soon after he was born, and he was cared for by his mother's sister because his father was old and infirm. He left school early and trained to be a carpenter. His daughter indicated that he had always had an interest in honey bees but that he only started serious beekeeping in 1908. At this time, he started to experiment with the introduction of queens into colonies of Cape honey bees and became interested in the laying workers that he observed in the colonies. Both he and de Villiers (1883) had observed that workers lay eggs, but it was Onions who undertook careful observational experiments that showed that the workers had activated ovaries and that their eggs gave rise to workers rather than drones. It is remarkable that this relatively unschooled beekeeper had discovered a phenomenon that all the "experts" of the period considered to be false:

> Mr G.W. Onions P.O. Retreat, Cape District, has deposited with the Government Entomologist an account of observations and experiments conducted by himself, which tend to show that laying workers of the native black honey bee are far more common than is generally supposed, and that their eggs generally produce workers and not infrequently queens. Mr Onions is a keen bee enthusiast who is well acquainted with the standard British and American publications on bees, and he believes that he has fully substantiated his astonishing statements. (Onions—*Agricultural Journal of the Cape of Good Hope,* 1909)

This extract is the first documented record of the discovery of thelytoky in the Cape honey bee and established Onions not only as an important figure in "African apicultural history," as suggested by Hepburn (1995), but also as a significant figure in apicultural history in general.

The significance of this discovery for the enrichment of our understanding of honey bee biology has taken almost a century to be fully appreciated.

[1] The biographical material presented here was sourced from the article by Hepburn (1995) in the *South African Bee Journal.*

6.9. The assassins and social parasites

Robbing the resources of a colony is a straightforward war between two colonies, with organized swarm raids of the attacking colony and fierce defense by the colony being robbed. Parasitic workers, however, follow a very individual route to take over not only the food resources but also the entire colony.

Worker reproduction is normally under strict control of the fatty acid queen pheromones produced in the mandibular glands. This is also the case in colonies of *A. m. capensis*. Queens of the Cape honey bee produce substantially more of the queen pheromones than do queens of other subspecies (Figure 6.12). This is necessary because workers of *A. m. capensis* require higher pheromone levels to refrain from activating their ovaries. So if a worker of *A. m. capensis* ends up in a colony of another subspecies, it will activate its ovaries, produce queen pheromones, and quickly become a false-queen. If such a false-queen meets the resident queen, a lethal fight ensues that is almost invariably won by the queen. However, the *capensis* workers have two behavioral tactics that allow them to survive. The most efficient one is simply to avoid the queen. The colony is large with a highly compartmentalized structure, with many combs providing ample opportunity to stay clear of the queen. Even in single-frame observation hives, we have shown together with Randall Hepburn that false-queen workers were more often found on the opposite side of the comb from the side with the queen (Moritz, Crewe, and Hepburn 2002). In large colonies, such behavior will allow the false-queen worker to stay in safety far away from the queen (Pirk, Neumann, and Hepburn 2002). Beekeepers are inadvertently often very kind to these false-queens and help them by placing a queen excluder grid in the colony. This is of course not to make the false-queen happy; it is to confine the queen to the brood nest in the bottom part of the hive and to allow only the workers to have access

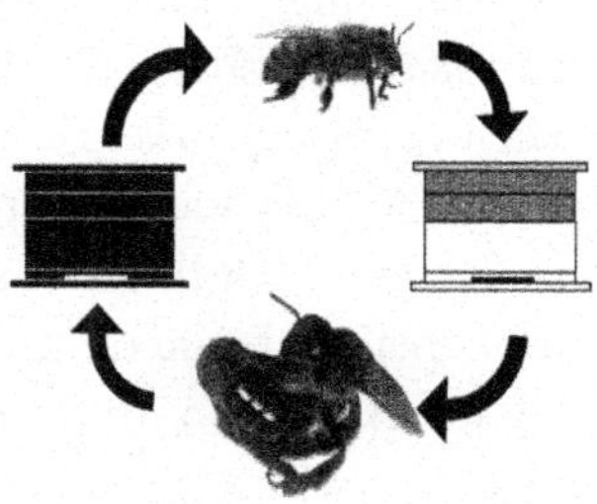

FIGURE 6.12 *Life cycle of the parasitic Cape honey bee. A parasitic bee enters the colony to establish a small brood nest (e.g., in the honey supers) where the queen has no access. The intruder or her offspring kill the queen (bottom) to take over the colony and use its workers to produce parasitic worker offspring exclusively. Eventually, there are no host workers to care for the brood, and the colony collapses and sends out parasitic workers to start the cycle of colony usurpation anew.*

to the supers to store the honey. However, this will allow the *A. m. capensis* worker to establish its own brood nest away from physical queen control. Once the first worker offspring of the false-queen emerges, a more critical phase starts. Because these worker are almost clonal offspring of the laying worker, they can also easily become false-queens. For them, it is more difficult to escape from the queen pheromones because with every new false-queen that develops, there are more sources of pheromones in the colony, making it practically impossible to escape all of them. Indeed, it appears as if workers compete to produce the strongest signal. These worker-produced queen pheromonal signals are not able to suppress the ovary activation of their clonal sister workers. Hence, there is now a large number of potential false-queens in the colony that will release aggressive conflict with the queen. So although the queen has a high chance of winning the fight, the probability for losing is not zero. With increasing numbers of parasitic workers, she will eventually lose a fight, and the colony then becomes queenless. However, the workers of the host colony will not engage in queen rearing because the queen pheromone levels are sufficiently high and the false-queen is accepted as the controller of the colony. Eventually, the host workers will die and will all be replaced by the offspring workers of the false-queen. Because these offspring workers are asexually produced with no sexual recombination and with crossing over strongly suppressed among their chromosomes, they are called pseudo-clone workers. They are almost genetically identical, and all share the parasitic trait. Thus, all of these workers will activate their ovaries to engage in egg laying, but there will be no brood rearing because no non-reproductive workers are present to perform normal colony activities (Hillesheim, Koeniger, and Moritz 1989). Thus, despite all the eggs, there will be no offspring because no brood care occurs in the colony. Eggs are laid and eaten but not reared to the adult stage. The colony will therefore rapidly dwindle once the host colony workers have died, and the parasitic workers can only survive if they find new host colonies in which to establish themselves. Colonies that do not have an *A. m. capensis* queen that can pheromonally suppress these workers can now serve as new hosts. The social parasites will invade a new colony, but the host colony will only die when the host workers can no longer provide brood care and have been replaced by the parasitic worker offspring of the invaders. Large-scale migratory beekeeping in South Africa facilitated the spread of *A. m. capensis* colonies out of their endemic range, resulting in the "capensis calamity," which caused a major disruption to commercial beekeeping operations and resulted in thousands of colony losses in apiculture in 1990 (Figure 6.13). A single parasitic clone turned out to be most virulent of all the potentially invasive clone offspring and was found to be responsible for the massive impact on beekeeping.

The genetic mechanism underlying this transition from a social worker bee to a social parasite can be understood in some detail through discussion of

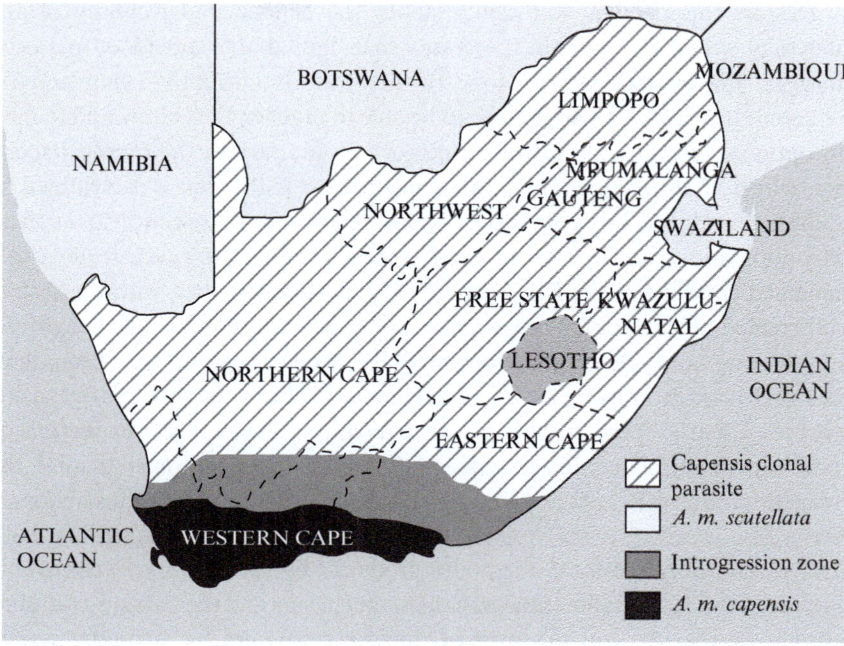

FIGURE 6.13 *Distribution map of the subspecies of honey bees in South Africa and that of the parasitic A. m. capensis worker clone (1990) that emerged as a result of the "capensis calamity" causing massive losses to South African apiculture.*

the following experiment. Queens were reared from laying worker-produced larvae and inseminated with a drone from arrhenotokous stock (Lattorff et al. 2007). The offspring hybrid queen was then backcrossed to a drone from the thelytokous stock. This queen produced workers with a clear-cut 1:1 segregation between those exhibiting thelytoky and those exhibiting arrhenotoky. A mapping study identified a genomic region of chromosome 13 that was tightly linked to the trait. Subsequent fine mapping suggested the honey bee homolog of the gene *gemini* is deeply involved in the trait. Differential splicing of two exons in *gemini* was shown to be associated not only with the mode of parthenogenesis but also with the activation of ovaries and production of queen-like secretions by the mandibular glands (Jarosch et al. 2011). Knockdown experiments induced ovary activation and queen-like mandibular gland secretions in arrhenotokous workers, which independently verified the results of the crossing experiments. However, a subsequent study using high-density genome mapping showed that although *gemini* is crucially involved in the control of the parasitic worker syndrome, it is not the genetic switch controlling the Mendelian segregation of parasitic and social workers. The first mapping study had shown us accidentally an important control gene but mislead us as to the actual switch mechanism.

Despite this detailed knowledge about the genetic underpinning of the thelytoky syndrome and the processes that caused the spread of parasitic lineages, the evolutionary biology of the trait is currently being actively explored. If it is such an advantage to be able to requeen the colony and to have an optional parasitic life history trajectory in addition to the classical social one, why does it not spread into other subspecies? If the trait is determined by a single major gene, this gene should be favored by natural selection and spread into populations under a broad variety of conditions. However, it obviously remained confined to the Cape region until humans meddled with the system and created a problem for beekeeping. The problem is confined to apiculture, and the wild honey bee populations in South Africa appear to have been unaffected. Obviously, transmission is important to facilitate the spread of parasitic workers. Clearly, the downside of the parasitic pathway is the extinction of the colony and the inability to requeen. So unless transmission is assured, the parasitic strains will quickly die out. Nevertheless, there is no obvious major selective disadvantage of thelytoky at the individual level, and the arrhentokous strategy can only prevail if the males produced by laying workers contribute more to overall population fitness than do the queens and the parasitic workers. In light of the frequent takeover of queenless colonies by intruding swarm propagules consisting of a few workers and a mated queen, queen replacement seems to have a very high fitness advantage indeed. But even within the endemic *A. m. capensis* population, the trait is not fixed, and both arrhentokous and thelytokous workers can coexist in the same colony. So the balance between individual- and colony-level selection will determine where and when thelytokous parasitic worker bees are successful.

{ 7 }

Diseases, Pests, and Parasites

FIGURE 7.1 *A female Varroa destructor mite (left) and Aethina tumida beetle (right), two of the major drivers of global colony losses in beekeeping (not to scale).*

7.1. Immunity to pathogens in honey bees

7.1.1. LOW NUMBER OF IMMUNE GENES

Honey bees are exceptionally clean, maintaining their nest cavity through a variety of tasks performed by the in-hive worker bees that contribute to nest hygiene and sanitation. They groom themselves whenever there is an opportunity, debris is removed from the colony, caretakers remove dead bees, and they defecate outside the nest. Inside the hive, the empty cells in the combs are polished until they are clean and shiny in preparation for the queen to lay eggs. Cells with honey stores are carefully sealed to ensure there is no leakage and to prevent the combs from becoming sticky. The stickiness of the honey is a major problem for honey bees. They are exceptionally vulnerable to honey contamination. Not only do sticky wings impede flight but also, even more important, honey can efficiently seal the tracheal spiracles, resulting in the suffocation of honey-contaminated bees. Hence, grooming is essential for colony life, and these strenuous efforts to maintain nest and body hygiene contribute to disease prevention. Alas, despite all this hygiene, honey bee colonies do occasionally get sick, and numerous infectious diseases, pests, and parasites are known to be major biotic factors affecting colony health and survival (Figure 7.1; Evans and Spivak 2010). The homeostatically regulated nest environment of 35°C and 60% relative humidity provides ideal conditions for bacterial and fungal growth that, coupled with the large throng of individuals in the colony, makes it easy for pathogens to spread within the colony (Schmid-Hempel 1998). So immunity is sure to be an important issue in preventing the spread of diseases.

Surprisingly, the honey bee genome revealed that the number of immune-related genes is very low compared to that of other insects (Honeybee Genome Sequencing Consortium 2006). Thus, despite the high risk of pathogen transmission in the colony, the genetically controlled innate immune competence is low. By comparison with other insects, honey bees lack a series of antimicrobial peptides (Evans et al. 2006), which seems puzzling given the huge diversity of pathogens known to infect them. The reduced individual immune capacity seems rather maladaptive at first sight—perhaps another dark side?

This reduced individual immune capacity is not a trait of the honey bees (genus *Apis*) alone but, rather, is common throughout the superfamily Apoidea, including both solitary and social bees. In the "10 bee" genome project, in which ten different bee species from different phylogenetic groups were screened for the immunity-related genes, the number of genes coding for immune-related proteins was low throughout all the species and no striking differences in the sizes of the different gene families were revealed (Kapheim et al. 2015).

7.1.2. ACQUISITION OF ANTIBIOTIC COMPOUNDS

Obviously, bees are still alive, with a remarkable number of species—approximately 29,000 throughout the world—and they seem to cope with pathogens in an appropriate and sustainable way despite having so few immune genes. So, how important is innate immunity for bees? One should keep in mind that all the bees in the 10 bee genome project exclusively forage on plant products. Indeed, almost all bees have a vegan diet, and this is important in relation to the risk of infection from food. All the plants they visit have one common problem: They need to be pollinated and therefore need to be attractive to bee pollinators. Just imagine you are one of those plants. You have evolved lovely colorful flowers to advertise your nectar. You know that bees will visit your flowers again if you are able to reward them with the nectar you offer. This revisiting exercise of the pollinator is an essential step for most flowering plants, particularly if they are self-sterile. So the pollinator needs to remember color and odor of the flower in order to provide a functional pollination service. If you now offer food in exchange for a service, you need to make sure it tastes good. If it is not sweet enough, the bees might simply ignore your other flowers and switch to another species of plant. Moreover, you would certainly ensure that your food is not pathogenic. Otherwise, the customer might simply die and never reappear. Bumblebees have been shown to identify and avoid flowers that are contaminated by a pathogen (Fouks

FIGURE 7.2 *The parasitic wasp Ammophila sabulosa struggling with a caterpillar. She will lay an egg in the prey where the larva will develop and can take full advantage of its host immune system. As a consequence, selection on its own immune system may have been relaxed.*

and Lattorff 2011). The nectar offered in a flower should be prevented from fermenting so as to remain attractive to pollinators. Providing these food conditions comes at a considerable cost to the plant. Most nectars are solutions with low sugar concentrations that are decidedly prone to fermentation if contaminated by bacteria or fungi. Hence, it would be highly adaptive if many nectars were to contain antimicrobial compounds that are secondary metabolites of the plant. Foraging bees would then collect these compounds while gathering nectar and pollen for food and could use them to prevent infections in the colony (Erler and Moritz 2015). So all bees, including honey bees, benefit from the food that is provided, but they also benefit from the antimicrobial compounds provided by the plants. Honey bees gain an added benefit if these compounds are stable because they can store them in the honey for long periods of time. In addition, honey bees collect propolis from plants, which has strong antiviral and antibiotic compounds that help in the treatment of various infections (Simone-Finstrom and Spivak 2010). The low number of immune genes in the honey bee genome (and also in all other bee genomes screened to date) may therefore be a consequence of the bees' specific diet that provides antimicrobial compounds from plants rather than a specific adaptation to a social life history. In addition, it must be kept in mind that phylogenetically bees are believed to have originated from ancestral parasitic crabronid wasps (Debevec, Cardinal, and Danforth 2012). These wasps are solitary parasitoids that lay their eggs in other insect hosts, and their larvae develop within the hosts (Figure 7.2). So they might be able to use a host's immune system throughout their development from egg to adult. They can also take up antimicrobial peptides produced by the host; therefore, any selection on their own immune system may have been relaxed.

7.1.3. NEST SANITATION

Honey bees make strenuous efforts to keep the hive as sterile as possible. The wax they produce for comb building is self-produced, sterile (Hepburn 1986;

FIGURE 7.3 *Workers of the stingless bee Trigona hypogea feeding on the tongue of a dead toad.*

Hepburn, Pirk, and Duangphakdee 2014), and the fatty acids present in it have been shown to have potent antibiotic activity (Gołębiowski et al. 2013). They line the nest cavity with propolis, which is composed of foraged plant resins with high antiviral and antibiotic potency. They forage on almost sterile plant products, and the only occasion when they might encounter a bacterial infection is when foraging for water. Compared to other social insects living in the soil and/or scavenging on cadavers, their foraging activities are notably sterile. Most infections will be transmitted by interactions with other honey bees, either after robbing other honey bee colonies or by drifting bees ending up in the wrong colony due to orientation errors. Testing the various hypotheses proposed for the reduced number of immune-related genes is clearly possible. If it is the vegan diet that drives the loss of immune genes, then the three carnivorous bee species of the genus *Trigona* that feed on carrion (Roubik 1982; Camargo and Roubik 1991) and also live insect larvae (Mateus and Noll 2004) will provide an excellent test case (Figure 7.3). The explosion in the availability of complete genome data will almost certainly provide the data that will distinguish which of the various theories for the reduction of immune genes, if any, remain plausible.

7.2. Social immunity

The reduced innate immune competence of honey bees has also been attributed to a so-called social immune system in addition to the innate one (Cremer, Armitage, and Schmid-Hempel 2007; Cotter and Kilner 2010). Although this has been shown to be untrue because the immune system was reduced well before sociality evolved (Kapheim et al. 2015), social interactions in the colony nevertheless play an important role in dealing with diseases and pests. A single infected worker may not be a problem for colony health if this worker is removed from the colony before it can transmit whatever diseases it has. Whereas solitary organisms try to confine a disease at the cellular level—for example, by apoptosis—a colony can do this at the level of individual colony members. This colony-level response was identified long before we knew about the cellular mechanisms regulating apoptosis.

Honey bee workers can handle infections at the colony level by specific adaptive behavior toward the infected individuals. Best known is "hygienic behavior," in which the workers uncap brood cells and inspect a larva or a pupa to determine whether it is infected with a disease (Rothenbuhler and Thompson 1956). Any diseased brood is then removed from the comb, and this trait has been shown to be most efficient against many brood diseases (Evans and Spivak 2010). What seems very adaptive, however, is arguably quite cruel. Diseased individuals are killed and eliminated from the colony—a clear case of systematic euthanasia.

A similar behavior can also be seen in the case of diploid drones (Mackensen 1951). These drones occur through inbreeding, since their egg carries the same sex allele as the male sperm. In these cases, the larvae are male, and they are removed from the hive within three days of hatching from the egg (Woyke 1969). Because the diploid drones are not fertile (they produce diploid sperm; Woyke 1980), this again seems to be highly adaptive. The diploid drones are eliminated at a very early stage of development, when the investment in them has been minimal. It seems very plausible: The sooner individuals who do not contribute positively to overall colony fitness are removed, the better because resources allocated to these colony members eventually end up as food for predators rather than benefitting the colony. This is a clear case of euthanasia in its darkest sense. Of course, it would have been even more efficient had the diploid drones not been produced in the first place. Rather than having the queen waste comb space, ovary activity, and protein resources, she should simply not produce diploid males. There seems to be little, if any, adaptive merit in using this genetic mechanism for sex determination when alternative, more efficient mechanisms are in place in other male haploid hymenopteran species (see Chapter 8).

Nevertheless, killing of diseased or disabled bees by healthy ones is not only directed toward brood. It can occasionally be observed at the adult level. Workers attack infected workers to remove them from the colony. If this is not possible, they do not hesitate to cut them into pieces to allow the pieces of the corpses to be carried out of the nest. The behavioral repertoire used against diseased nestmates can by highly variable. Whereas some workers are actively aggressive to infected individuals, others simply avoid them, thereby reducing transmission in the colony.

This became very clear in a series of simple experiments in which we observed the behavior of healthy workers toward a *Nosema*-infected and a sham-treated control worker bee (Biganski et al. 2018). In a first series of experiments, we observed highly aggressive interactions by the healthy workers directed toward the infected ones. As a consequence, the infected bees were right at the center of the social network with many recorded interactions. The same experimental setting produced the opposite result in a replicate study using workers from a different colony at a different time of the year. Now the infected bees became very peripheral to the social network, reflecting the isolation of these bees from the social group. So the response to infected nest members with the same disease can be completely bipolar: either intensive interaction or social isolation. Depending on the disease, neither of these responses may be sufficiently efficient to eradicate the pathogen from the colony. Removing seemingly redundant, contagious, or dead individuals from the nest is a well-known behavior of honey bees (Figure 7.4). After the mating season is over, the residual drones that have not successfully mated with a queen are forcibly evicted from the colony. The workers employ very drastic measures against unwanted colony members, although identifying which cues are used to release this intracolonial aggression against the offending individuals is a wide-open field of research. The aggression against the drones is particularly puzzling because the drones remain unchanged, but external seasonal factors appear to cause a fundamental change in worker behavior toward them (Boes 2010). Drones that have been readily accepted and embraced in the hive suddenly release aggression and are evicted from the nest.

7.2.1. ABANDONING THE NEST SITE

The most extreme way in which a colony can get rid of a brood disease is by completely removing all the brood in one stroke. The colony does not do this by removing the brood one by one from the cells; rather, it simply abandons the old nest site altogether, leaving the brood behind. Hence, the adults of the colony simply swarm off in search of a new nest site. The cost to the colony is huge because thousands of undeveloped colony members are left behind and are generally eaten by a variety of vertebrate or arthropod predators. This is colonial filicide.

Although the cost is high and undoubtedly cruel to the developing offspring, this swarming behavior seems to pay off at the colony level. The adults simply abscond to start a new nest at a presumably better site without the

FIGURE 7.4 *Dead (but still infective?) worker bee being carried away from the colony by an undertaker worker.*

burden of an infested or infected brood. This disease-driven absconding is very characteristic of tropical honey bee colonies. For these colonies, even seemingly minor disturbances can cause nest abandonment. This extraordinary behavior is considered to be a major driver of brood disease resistance in African and Asian tropical and subtropical honey bees. Although the behavior may rescue the colony, it is a bane in the lives of beekeepers. For the beekeeper, absconding results in colony losses from the apiary even though the colonies actually survive (just elsewhere). Today, we know that all honey bees south of the Sahara are resistant or tolerant against the most virulent brood diseases, including American foul brood and infestation with *Varroa destructor*. Honey bees in temperate climates cannot easily follow this path of nest abandonment. They need to prepare for long winter phases that require both large food stores and large colony sizes to maintain colony temperature. Absconding can only be used under extreme conditions such as floods or fires that would otherwise destroy the entire colony. Beekeepers value this more conservative life history pattern, with low colonizing capacity but strong investment in the status quo—resulting in more stable numbers of colonies in apiaries. The bees remain in the same place, which is obviously advantageous for beekeeping, but the cost is the loss of an important natural mechanism of ridding colonies of brood diseases. One way out of this dilemma is the establishment of artificial swarms by the beekeeper, which is an apicultural alternative to chemical treatment to rescue the colony. Here, the beekeeper destroys the infected brood and starts a colony with the adult bees at a new location in a sterilized hive.

7.2.2. SELF-MEDICATION

Honey bee workers are not always cruel to diseased nestmates—actually only occasionally. On other occasions, they actively provide therapeutic medication by feeding antimicrobial compounds to their nestmates. For example, food jelly fed to both larvae and adults contains many antibiotic compounds, including the antimicrobial peptide royalisin, a variant of honey bee defensin-1, as well as various fatty acids that inhibit bacterial growth (Fujiwara et al. 1990). Honey is also known to have considerable antibiotic properties. This is not just because of the high sugar concentration (>80%), which results in a high osmotic pressure and prevents bacterial growth in the stores. Even when diluted, honey maintains its antibiotic properties (Erler et al. 2014). This is not surprising if we examine the food chain of honey bees. They feed on nectar, which has a low sugar concentration and high water content. Hence, the nectar is prone to fermentation, unless the plants prevent fermentation of the nectar through the addition of secondary plant compounds. The antibiotic properties of diluted honey are used in the field of "apitherapy" but have their origin in the context of disease prevention in the hive because the larva are fed with honey diluted with food jelly containing considerable amounts of water.

Although apitherapy arguably reflects the dark side of biomedical science, often through the lack of rigorous experimental control in testing the efficacy of its products, the secondary plant metabolites in honey clearly do have beneficial health effects on honey bees. In laboratory assays, worker bees seem to be able to treat themselves: Workers infected with *Nosema ceranae* chose a different honey from that of uninfected individuals in a laboratory dual-choice test (Gherman et al. 2014). This in itself would not have been remarkable had they not chosen the honey with the higher antimicrobial properties and consequently reduced the level of their *Nosema* infection. One might argue that such laboratory experiments are not particularly decisive and prone to experimental artifacts. However, the point of the study was not to show that this is an important mode of self-medication within the hive, but given that honey bee workers are able to discriminate between two honeys in a glass Y maze with different effects on a pathogen, there is the potential that they could use this behavior in an adaptive way. This form of self-medication has still to be demonstrated inside the hive, and until then, this remains an appealing theory that may easily be proven wrong if the honey bees decide not to make full use of their skill. Even if they do use the ability to discriminate among different honey sources in the hive, we would clearly expect considerable genetic variation among workers similar to that shown for behavioral task specialization among the workers based on the subfamily to which they belong. This has been shown to be a consequence of genetic variance resulting from the polyandrous mating system of the queen. It would be very surprising if all workers were to perform this complex task equally well.

7.3. Dealing with unwelcome visitors

7.3.1. PUTTING PESTS IN PRISON

One of the most curios adaptations of honey bees to a parasite may be the interactions between workers and the small hive beetle *Aethina tumida* (see Figure 7.1). The beetle is a common parasite in African *Apis mellifera* subspecies but can have devastating effects on European subspecies. In the United States, it has caused dramatic colony losses. These losses have elicited grave concern among European beekeepers because the parasite has recently been detected in Italy and is currently spreading. African honey bees have a series of adaptations that allow them to mitigate the effects of an infestation of the colony by the beetles. A key adaptation is the propensity to abscond, leaving the old nest site to the parasites and starting anew elsewhere. However, there are also less radical adaptations. Some workers actually feed the beetles rather than fight them, which seems to be a fundamental maladaptation to a parasite. Yet, it is not uncommon that parasites of social insects elicit trophallactic behavior in the host. By emitting chemical and/or tactile signals, the parasites can release social behavior in the host that is otherwise only directed toward colony members. However, not all workers feed the beetles; there are also others that chase the beetles around the hive and eventually herd them into inaccessible cracks and corners of the nest. After having cornered the beetles, the worker bees construct propolis walls around them, and guarding bees ensure that they do not leave their "prison" (Neumann et al. 2001). One might expect that the workers would proceed to entomb the beetles in propolis and simply let them die in place. However, this is not what happens. The workers are not that cruel and do not simply starve the parasites to death. Instead, they leave small windows in the prison walls—small enough so that the beetles cannot easily escape but large enough to feed them in the prison (Ellis et al. 2002; Figure 7.5).

Clearly, it would be much more efficient to simply seal the prisons off and ignore the beetles trapped within them. This is what workers do if they encounter mice or other large intruders that they cannot remove from the colony. They simply entomb the intruder in thick layers of propolis that prevents microbial decay due to its potent antimicrobial compounds. The beetles seem to be able to trick some of the workers but not all, and again it is not known why this is adaptive. The beetles may simply exploit some other more important adaptation from which they benefit and elicit trophallactic behavior from worker bees (Neumann et al. 2015). Again, stingless bees (*Trigona carbonaria*) appear to handle small hive beetles much more efficiently by immediately coating invading beetles with a mix of resin, wax, and mud (Greco et al. 2010): No fuss—clean kill—end of story.

7.3.2. PREVENTING REPRODUCTION OF PARASITES

It appears that honey bees do get by in their struggle with pest and pathogens and also that natural selection can be extremely efficient with regard to surviving parasite pressure. The best example may be the resistance of honey bee colonies to the parasitic mite *V. destructor* (see Figure 7.1). The *Varroa* mite parasitizes both brood and adults by feeding on hemolymph, while its reproduction is limited to sealed brood cells during the period that the bees are pupating. The female mite enters the cell just before capping to lay first a haploid male egg and subsequently female eggs. The single male then mates with his sisters and

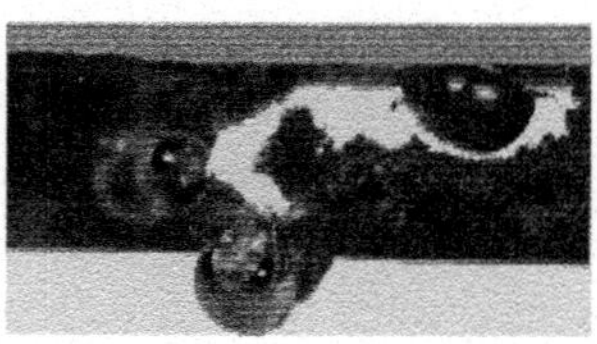

FIGURE 7.5 *Workers guarding and feeding a small hive beetle in a propolis "prison."*

dies when the adult bee emerges together with the mated female offspring and the mother mite (Rosenkranz, Aumeier, and Ziegelmann 2010). This mating system results in high inbreeding in the mite population.

In the eastern honey bee, *Apis cerana*, which is the original host of *Varroa* mites, the mites do not cause much damage and the bees are apparently tolerant of the infestations. Tolerance in *A. cerana* is attributed to a variety of factors, including grooming behavior, killing mites by cutting off their legs and chelicerae that are then found on the bottom board of hives, and the lack of reproduction in worker brood. The mites only reproduce in the drone brood and hence only during the short period of the season when the colony produces drones. Due to this efficient resistance, it is exceptionally difficult to find *Varroa* in an *A. cerana* hive unless there is drone brood in the colony (Boot et al. 1999). A most peculiar mode of preventing *Varroa* from reproducing in worker brood has been suggested by Page et al. (2016). They noted that *Varroa* infestations often caused *A. cerana* worker larvae to delay development or die in the brood cell. Indeed, in combination with hygienic behavior, this would prevent *Varroa* from reproducing, which is why they termed this "social apoptosis." Puncturing larvae with a sterile needle caused much more lethal damage to *A. cerana* worker larvae than to those of *A. mellifera*, which may indicate that the larvae of *A. cerana* are more fragile than those of *A. mellifera*. However, the conclusion that this is a mechanism resembling adaptive apoptosis of higher organisms may be premature. If this mechanism were to be a major driver of *A. cerana* tolerance to *Varroa* infestations, it should be absent in drone larvae where *Varroa* reproduces in *A. cerana* colonies. Until this is tested, the absence of reproduction in larvae may be an interesting observation but with only limited impact on establishing a stable host–parasite equilibrium between *A. cerana* and *Varroa* in Asia, and other mechanisms may be more important for the tolerance of *A. cerana* to the *Varroa* mite.

Unfortunately, the mites were able to exploit a new host in Asia and infect the imported western honey bees, *A. mellifera*, where they could reproduce in worker brood. This caused a major problem because mites could reproduce throughout the year and were not confined to the period when drones are produced by the colony. The mite population in colonies increased to the point where their numbers were lethal to the colony. If not treated with acaricides, colonies died within three years after a first infestation. However, not all colonies died. Surprisingly, in regions with high densities of feral or wild colonies, *A. mellifera* colonies seemed to get by. In particular, in Africa south of the Sahara, wild populations seemed to be almost unaffected and colonies survived well without any treatment. In Europe, several populations were identified in which colonies survived *Varroa* infestations without treatment (Kurze et al. 2016). In two of these, one in Sweden and one in France, the reproduction of mites in the brood cells seemed to be inhibited. In both cases, the resistance to the parasite was based on inhibition of the mite's reproduction, with the mites entering the brood cells but failing to produce offspring (Le Conte et al. 2007; Locke and Fries 2011). Many mites produced no offspring, not only in worker brood but also in drone brood. This resistance to mite reproduction could be used to search for specific genes that would interfere with their reproduction (Behrens et al. 2011). The mite reproductive cycle is extremely fine tuned to the larval development of the host. There is only a very narrow time window of approximately 12 hours that the mite can use to activate its ovaries (Figure 7.6). If she misses this opportunity, she will remain infertile and confined in the cell until the worker emerges. Clearly, the reproductive physiology of *Varroa* is very closely regulated by host signals, providing a potential lever for the selection of true resistance if the host could simply modify such a signal. Other mechanisms also interfere with mite reproduction, including the so-called *Varroa* sensitive hygiene (VSH) lineage that has been bred to produce workers that remove parasitized pupae from the cells (Harbo et al. 2009), which may be similar in outcome to the "social apoptosis" mechanism observed in *A. cerana* (Page et al. 2016). However, whenever natural selection has been allowed to operate (i.e., no interference by bee breeders), it seems that direct interference with mite reproduction becomes a frequent phenotype in the honey bee population (Le Conte et al. 2007; Locke and Fries 2011; Strauss et al. 2016). If this is true, it would be a most elegant way for the honey bees to deal with the problem, and the dark side would actually arise from all the beekeepers who keep treating their colonies with acaricides rather than allowing natural selection to take its course (see also Chapter 9.

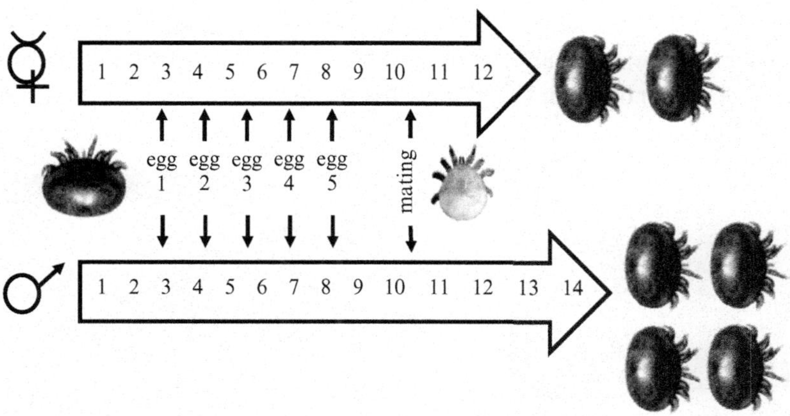

FIGURE 7.6 *Reproductive cycle of V. destructor during post-capping of A. mellifera brood. The female enters a brood cell just before capping. The first egg laid develops into a male (white individual to the right), which sequentially mates with its sisters before they emerge as fecund mites ready to infect new cells. (Top) Development in a worker brood cell results in only two offspring mites. (Bottom) Development in a drone brood cell allows for the production of four offspring mites due to the longer developmental time.*

7.3.3. ATTACKING GIANTS

Hornets are a rather diverse group of large wasps of the genus *Vespa*. They are keen generalist predators but can also hunt honey bees, which is why beekeepers are not very fond of them. Nevertheless, in some countries, hornets are on the red list of endangered species and enjoy conservation protection by law. Some species are more dangerous to honey bee colonies than others. The first reports of hornets eliminating complete honey bee colonies came from Japan, where *Vespa mandarinia* was considered a lethal threat to *A. mellifera* colonies (Matsuura and Sakagami 1973). This is a sad story resulting from the introduction of the western honey bee *A. mellifera* into Asia for honey production. *Vespa mandarinia* is a generalist predator species but can attack and completely destroy colonies of other social wasps and bees, including honey bees. The endemic Asian *A. cerana* can defend its colony efficiently by balling the hornets and killing them with heat (Figure 7.7). The introduced *A. mellifera* is much less efficient in this regard and often completely helpless. Beekeepers have to construct specific flight entrance devices to keep the loss of colonies due to *V. mandarinia* predation at bay. Although all hornet species can prey on adult bees, the attacks of *V. mandarinia* are considered the most dangerous because they can kill the entire colony. Although these hornets hunt adult honey bee workers, they prefer the brood in the colony and have evolved a highly specific raiding tactic to overwhelm defending honey bees. In order to get access to the brood, they first test colonies' defensive response to identify those that are sufficiently weak in a so-called hunting phase. In this initial phase, the colony damage is limited. The wasps will catch defending workers, but if there are persistent attempts to ball the wasps and heat them up, they will eventually withdraw to seek another victim. Once a colony with poor defenses is identified, the behavior of the wasps changes dramatically. They will now focus their efforts on these colonies and switch to the so-called slaughter phase. Any worker caught

FIGURE 7.7 *(Left) Vespa mandarinia facing a tight cluster of honey bees. (Right) Heat balling of the unlucky attacker by A. cerana honey bee workers.*

will not be taken as food to the wasp nest. The corpses of dead bees are now simply dropped in front of the bee colony as soon as the bees have been killed. It takes a wasp approximately 10 seconds to kill a bee. Matsuura and Sakagami reported that 30,000 honey bees were killed by only 30 hornets within three hours. Only two hornets were killed by the bees during this battle. Once all defending workers have been killed, the hornets enter the colony to first rob the brood and only then feed on the corpses. During the occupation phase, hornet workers carefully guard the bounty and defend the occupied honey bee colony against others, including other hornets and any other potential predators, such as humans and vertebrate animals that may be unaware of the situation prevailing in the hive.

Clearly, despite more than 100 years of beekeeping with *A. mellifera* in Asia, the introduced honey bee has not evolved any suitable defense mechanism against this predatory wasp. It is therefore of some concern that *Vespa velutina*, another Asian wasp species, has recently been introduced to southern Europe (Monceau et al. 2014). Although it has been identified as an additional problem for apiculture, its attacks are fortunately not as lethal as those of *V. mandarinia* because they primarily prey on honey bees outside the colony. Nevertheless, it adds to the difficulties beekeepers must deal with, including pests, pathogens, and poisoning.

Apis mellifera seems to be defenseless against *V. mandarinia* and requires the help of beekeepers to ensure survival. Creative Japanese apiculturists have designed various traps that can be attached to the hive to assist *A. mellifera* colonies to defend themselves against these predators. The traps use a wire mesh sufficiently narrow such that the wasps cannot pass through it but that allows the much smaller honey bee workers to pass through. In addition, the traps catch wasps that try to flee in response to honey bee attacks. Because their escape route is generally upwards, a funnel leads the wasps into a cage out of which they cannot escape, but the honey bees can escape. The beekeeper can then collect and kill the wasps caught in the trap. Although this procedure prevents the complete slaughter of the colony, it is not effective in reducing the loss of honey bee workers during the hunting phase. The wasps simply sit and wait outside the trap to catch the workers, so destroying hornet nests in the vicinity of apiaries becomes the only remaining defense.

{ 8 }

The Idiosyncrasies of Sex and Reproduction

> If, for instance, to take an extreme case, men were reared under precisely the same conditions as hive-bees, there can hardly be a doubt that our unmarried females would, like the worker-bees, think it a sacred duty to kill their brothers, and mothers would strive to kill their fertile daughters; and no one would think of interfering.
>
> —Charles Darwin (1890, p. 99)

Knowledge of honey bee sex and mating, although the subject of extensive speculation and reinvention, dates back to Aristotle (350 BCE), who informs us that "others, again, assert that these insects copulate, and that the drones are male and the bees female" (*The History of Animals*, Book 5 part 21). Although reproduction in the honey bee is possible without sex (males are produced in this manner), queens have to mate with males in order to lay fertilized eggs and produce female offspring. Reproduction in the honey bee is not simply an issue of individual reproduction but, more significantly, it is about the generation of new colonies to achieve larger populations. Colony fission through reproductive swarming is the basis for this population growth. This form of reproduction is not common in social insects with large colonies. Only a few of the more advanced social hymenopteran species pursue this pattern of reproduction. The queen does not found the colony on her own but, rather, with assistance of the worker force (termed "swarm founding"). In the honey bees, this form of reproduction leads to a series of paradoxes that can be illuminating about the degree to which evolutionary solutions are based on what Stephen Gould (1981) called "the principle of imperfection."

8.1. The swarming paradox

Reproductive swarming is a process that humans have observed since they first started handling bee colonies in prehistoric times (Figure 8.1). A passing honey bee swarm is indeed most impressive, and it was considered in some cultures to have ritual power.

FIGURE 8.1 *Cave painting depicting a human interacting with a swarm taking off (Harrismith, South Africa).*

Understanding the biological processes that define the behavior of swarms had to wait for the arrival of the 20th century with the pioneering work of Martin Lindauer (1955) and subsequently that of Thomas Seeley and co-workers, who conducted a series of painstaking behavioral observations and experimental studies that meticulously dissected the various steps in the swarming process. Seeley (2010) marvelously reviewed the biology of swarming in his monograph, *Honeybee Democracy*, demonstrating the full intricacies of the underlying behavioral processes. We now have a rich understanding of the way in which the seemingly confused cloud of bees flies from one location to another without losing substantial numbers of its members.

In brief, preparations to issue a swarm begin about a week before the actual swarm takes off. The old queen stops laying eggs, reduces the size of her ovaries, and loses weight, allowing her to regain her ability to fly. The workers start rearing new queens, and before these emerge the first swarm takes off. If the queen does not join the swarm, the bees will return to their home colony. If the queen joins them, then they will leave the colony to cluster at a bivouac resting site while scouts start searching for and eventually agree upon a suitable new nest site. Swarming has an immediate impact on the cognitive processes of the individual bees within it. After having learned the location of the old nesting site for weeks, they completely forget about it and will not return to the old place. The departure of the swarm from its bivouac is then initiated by the scouts—interestingly, the very same workers that made the swarm leave the hive in the first place (Rangel and Seeley 2008; Rangel, Griffin, and Seeley 2010). They produce a series of peculiar sounds, including a piping signal that serves as a primer for departure. Next, the same bees are the first to start buzz-runs (*schwirrläufer*; Lindauer 1955), which provides the signal that releases the departure of the swarm. This buzzing sound was first quantified by Esch (1967) and later confirmed by a long series of follow-up studies further clarifying the details (reviewed in Seeley 2010; Schlegel, Visscher, and Seeley 2012). Guided by a small number of scout bees that perform streaking flights through the swarm, the bees will eventually arrive and settle into the new nest cavity, which is facilitated by the release of queen and worker pheromones. The search for the new nest site seems to be a particularly efficient and highly concerted effort given that thousands of workers need to agree to fly to the same place—an obvious masterpiece of communication in which a very small proportion of the workers are able to guide the colony to its new location (Beekman, Fathke, and Seeley 2006).

8.1.1. THE OLD QUEEN SWARMS

Communication during swarming may be an example of one of the most highly skilled behavioral adaptations of honey bees, as is the resulting colony fission that seems to be reasonably optimized to achieve its purpose (Rangel, Reeve, and Seeley 2013). However, from a more strategic and evolutionary perspective, the role the queen plays in the process and the actual departure of the swarm from the old nest site seem to leave a lot of room for adaptive improvement. Indeed, during this phase, the entire process appears to be highly maladaptive and leaves the naive observer rather puzzled. The first problem arises because the old queen flies off with half of the colony. This is unique in eusocial flying insects because the general rule is that the old queen remains and the young queen flies off to establish a new colony (Cronin et al. 2013). This seems to make evolutionary sense. The queen of large eusocial colonies must have large ovaries for the production of sufficient numbers of eggs to ensure colony growth. Most queens are therefore physogastric with a huge abdomen that makes them fairly immobile and precludes flight. Yet the honey bee queen needs to fly in order for swarming to take place, and she therefore has to reduce the size of her ovaries prior to swarming in order to regain the ability to fly. The queen necessarily stops laying eggs in preparation for the swarming flight several days before the swarm leaves. Yet reproductive swarming takes place at the peak of the foraging season when the queen can lay up to 2,000 eggs per day.

Because swarm preparation starts about a week before the actual swarm is issued, the colony is deprived of well over 10,000 young workers since the queen stops laying eggs. In an average colony, this is approximately one-fourth of the total worker force, so the price paid for swarming is high. Other bees

FIGURE 8.2 *A physiogastric queen of the stingless bee Melipona beechii inspecting an open cell before oviposition. This queen has lost the ability to fly, in contrast to the honey bee queen, which can regain its flight ability after suspending oviposition prior to issuing of the primary reproductive swarm.*

do not seem to burden their queens with such a demanding slimming exercise. Queens of stingless bees, the sister group of the honey bees and also highly social bees with colony fission, do seem to bother. They remain fully fertile and physogastric, and they simply will not and cannot fly after having started egg production (Figure 8.2). In stingless bee colonies the queen does not stop producing offspring even during times of colony fission because it is the young queens that leave the colony and not the old one. Hence, the question is not just about the honey bee queen stopping egg laying before swarming; it is also about why she does not develop physogastry like her stingless bee counterparts. A honey bee queen with much larger ovaries might be able to produce even more eggs per unit time and might enhance overall colony fitness if she did not need to fly during swarming.

However, physogastry has a clear downside: Laying many eggs is highly relevant to the fitness of the colony, but the ability to fly may be convenient because it allows the colony to migrate and abscond in cases of emergency and take the queen with it. So with regard to physogastry, there may be a trade-off between queen fecundity and colony mobility. Mobility seems to be a major adaptive advantage for honey bees that are specialists in foraging on mass nectar flows. Once a rich food source has been exploited and the season becomes unfavorable, it may be highly advantageous to move to a new place. Migratory swarms are well known in most honey bee species. The most extreme cases are the migratory swarms of the giant honey bee *Apis dorsata* in Asia that have been reported to travel hundreds of kilometers following rewarding honey flows.

Nevertheless, putting the incumbent queen of a successful colony at extreme risk seems odd. Leaving the shelter of the nest site exposes her to predators, pests, and parasites. Moreover, swarming occurs at a time with abundant food, and she needs to stop laying eggs just when feeding conditions are excellent. Although she will reduce the size of her ovaries prior to swarming, she may simply not regain sufficient flying ability and be lost in flight. So there is active selection against physogastry in honey bee queens, which is not necessarily adaptive for a queen of a fixed perennial insect colony. Wouldn't it be much more efficient for young queens to leave the hive with the swarm of workers instead of the old queen? Wouldn't it be better to swarm after the virgin queen has been mated so that she can start egg laying immediately after the swarm has settled at the new nesting site, as in stingless bees? If the huge number of stingless bee species can do this, why can't the depauperate number of extant honey bees species?

There is so much uncertainty about the future performance of the virgin queen that it seems a foolish gamble to hand over the colony to her. It is unclear at this stage whether the queen's royal daughter will mate and, if so, with which males (a parental concern in many societies), whether she will be able to maintain a colony that can issue a reproductive swarm in the next season, and what venereal diseases she will pick up during mating. So many uncertainties are associated with the establishment of a new queen, whereas the old queen has proven her quality for successful colony reproduction. The resources of the old nest are unnecessarily exposed to a high risk with the new queen taking them over. If mating is risky, the reproductive value of the old queen will be higher than that of a virgin queen, and hence she should be less exposed than the young queen. Of course, the old queen will eventually show signs of senescence. The colony, however, is well prepared should this happen. Once the old queen starts showing signs of low reproductive quality (e.g., sperm depletion or old age), she is replaced by the mechanisms governed by queen supersedure. A new queen will be raised by the workers in the presence of the old one. During supersedure, the new queen and the old queen can even coexist for a considerable time until eventually the old one disappears, either as a result of natural death or as a result of antagonistic behavior of the new queen or workers (Deyel 1980). So whereas supersedure seems to be an evolutionary rational mechanism, it is surprising that reproductive swarming based on swarming with old queens cleared the hurdle of natural selection as a functional mechanism.

8.1.2. HOUSE HUNTING

Another confusing aspect of reproductive behavior in honey bee colonies is the process of moving to a new nest site. The swarm issues before any house hunting has been done—that is, the swarm leaves the colony in search of an appealing tree branch or any other suitable structure where it can sit for a few days. Only then do the scouts start searching for appropriate cavities. Quite often they fail to find suitable nesting sites, and in the worst case this can result in the death of the swarm due to its exposed position during winter. This strategy is rather like packing the moving van, parking it in a no parking zone, and hoping to find a suitable home before the parking tickets accumulate. So quite clearly, a strategy of first looking for an appropriate site, then preparing the site for the queen, and only calling her when all is ready would seem to be far more efficient. This solution is not confined to the way people arrange to move from one home to another but also how most other social bees arrange their swarming processes, as can be seen in colony fission of stingless bees. Here, the workers identify a new site, transport resources from the old nest to the new one, start building combs, and only when the nest is well prepared is the queen guided to the new place. Also—hardly surprising—it is the young mated queen that leaves the hive, whereas the old physogastric queen that cannot fly because of her large and heavy ovaries remains behind. Behavioral ecology is a discipline renowned for elegant post hoc adaptive explanations for behavioral peculiarities, but in this case there seems to be no plausible explanation that helps the honey bee colony look well organized. They have evolved excellent skills in identifying new and suitable nest locations. The workers can even all agree on which is the best new nesting site based on a complex and highly sophisticated communication system that is beautifully described by Thomas Seeley (2010). However, it seems all of these remarkable house-hunting skills are used to support a highly risky reproductive swarming strategy.

8.2. The queen rearing paradox

We have already dealt in some detail with the process of queen rearing in preparation for swarming and emergency queen rearing. The colony will try to rear many queens at the same time, which seems perfectly sound to ensure that there are a few spare in case one fails. Indeed, the probability of the loss of a virgin queen can be high depending on environmental conditions. They can be lost due to predation, bad weather, and orientation errors; killed by invading parasitic workers; or squished between combs by the negligent apprentice beekeeper or bee researcher. So there are manifold reasons why it seems prudent for the colony to rear several queens. But then, after this hopeful start, there is a break in the logic and all the sound plans are overturned. As soon as the first queen emerges and while the others merrily peep in their cells so as to draw the attention of the queen that emerged first, she immediately proceeds to skillfully kill them all. This leaves the prudent observer rather speechless. Clearly, monogyny is the desired state of the colony, but why does the virgin queen kill them all before she is mated? This is certainly neither in the interest of the colony nor in her own interest because she would probably gain more inclusive fitness from a fully fertile sister taking over rather than herself should this be dictated by circumstances. It seems a rather bold decision to assume that mating will proceed without a hitch and to therefore kill all competitors as soon as possible. What if she meets a bee-eater on her mating flight? What if there are no drones around and the weather does not permit mating to take place? She would then only produce males before the colony collapsed due to a lack of workers. Drone-laying queens are also quickly eliminated by the workers, which will start oviposition themselves. Although it may certainly be in the interest of the queen to make sure she has no competitors in the colony, her inclusive fitness would be higher if she killed the other queens after her own reproductive value had been partially ensured by a successful mating. And yes, some honey bee queens seem to be wise. Queens of *Apis mellifera intermissa,* a subspecies in northern Africa, follow this seemingly more adaptive pattern (Hepburn and Radloff 1998). Many queens coexist in the colony, but only until the first one returns from the successful mating flight. When she comes back, the other queens are killed and monogyny is restored. So again, if *A. m. intermissa* can do this, why has this strategy not spread more widely? Why do the other subspecies run this seemingly completely unnecessary risk of colony loss? Even if the mating risk of a queen is low, it is never zero, and yet postponing the killing of sister queens would apparently come at little cost, if any.

The swarming process, one of the most spectacular events in honey bee biology, is far from being plausibly explained. We believe that honey bee swarming is anything but an optimal adaptation. It is probably one of those cases in which evolution has "jury-rigged" a solution for a difficult problem but failed to find a good one. It is a good example to show that evolution in biological systems is not so much about optimal solutions but more about feasible solutions that result in stable systems. The colony somehow gets by despite the legacy of behavioral repertoires from the good old days of solitary life and individual Darwinian fitness.

8.3. The sex locus paradox

The evolution of sex, one of the major transitions in evolution, is still far from being well understood. The costs associated with sexual recombination seem to be too high, and the fitness benefits seem to be too low. But this is a general problem addressed in many monographs by outstanding scholars (Maynard Smith 1978; Hartfield and Keightley 2012) and not specific to honey bees. Honey bees have their own specific problem. Sex in the honey bee is determined by ploidy. Males are haploid and females are diploid. Although this may seem odd to readers familiar with chromosomal sex determination with X and Y chromosomes, male haploidy is actually quite common in many arthropod orders and seems to be the rule for Hymenoptera. Indeed, evolutionary theory can convincingly show that male haploidy is actually a superior evolutionary strategy to that of having both sexes diploid (Kidner and Moritz 2016). One obvious advantage is the lack of any genetic load in male haploid populations because any deleterious recessive allele will be directly selected against in the haploid individuals. Male haploidy is thus actually a most promising invention and certainly an evolutionary stable strategy with profound evolutionary success. The question is rather: Why are there so many species with diploid males?

However, the sex determination paradox in honey bees arises not from male haploidy per se but, rather, from the way in which it manifests itself. For whatever reason, sex is determined by a single locus in the honey bee and in many other male haploid species as well (van Wilgenburg, Driessen, and Beukeboom 2006). Originally, this locus was termed the sex locus (Whiting 1943) before it was called the complementary sex determiner (*csd*; Beye et al. 2003) to better illustrate the mode of operation. Whenever an individual is heterozygous at *csd*, it will turn into a female. If it is hemizygous, as in the haploid drone, it will turn into the male sex. It will, however, also turn an individual into a male if it is homozygous. Hence, any diploid individual homozygous at the sex locus will turn into a diploid male. This happens if a queen's eggs carry the same sex allele as the fertilizing sperm of the male (Mackensen 1951). Diploid males, however, are not good news for the colony because they produce diploid sperm, causing triploid and hence lethal offspring. The workers have evolved a behavior to solve this problem post hoc by not investing in these sterile males. They quickly cannibalize diploid drone larvae after the larvae emerge from the egg (Woyke 1965, 1969). However, this nevertheless imposes a cost to colony fitness. In the extreme case, 50% of the eggs laid by the queen will be killed as larvae if mated with a single drone sharing one of her alleles at *csd*. One might argue that this is fine because the colony obviously gets by. Although it is true that the colony gets by, it is anything but fine. As Ross Crozier (1971) noted, using a single locus for sex determination may not be the most efficient way to determine sex in male haploid species. Indeed, we now know of various hymenopteran species that use more than one locus to determine sex (Crozier 1981, de Boer et al. 2015). In many bee species, diploid males are notoriously difficult to find and are extremely rare (Soro et al 2017). One reason may be that they are lethal and never reach the adult phase; however, another reason may be that other bees simply have a two-locus or multilocus system ensuring that diploid males are rare even after inbreeding. If not just one but several *csd* loci need to be hemi- or homozygous at the same time, the frequency of diploid males will be substantially reduced even in brother–sister matings and decrease exponentially with the number of loci involved (Figure 8.3).

If diploid drones are lethal, they are a substantial fitness cost and natural selection should favor any mechanism reducing their frequency. Multiple loci are obviously a most efficient way to reduce their frequency. The honey bee, arguably a rather derived hymenopteran species, must have had many phylogenetic opportunities during its evolution to duplicate sex determining loci and hence reduce the genetic load associated with the sex locus. So why didn't evolution

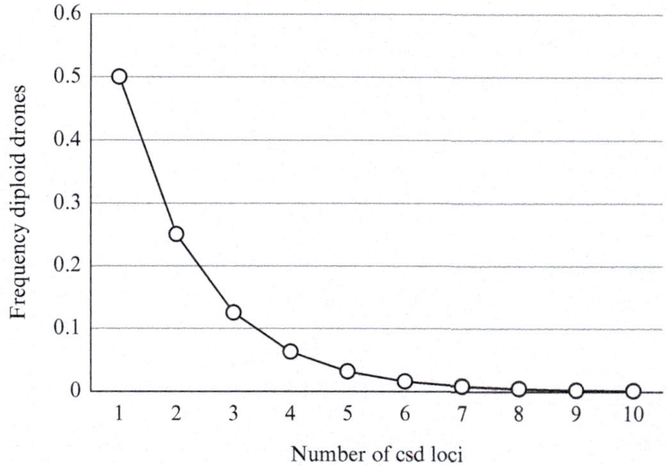

FIGURE 8.3 *The frequency of diploid drones (in a mother queen–son drone mating) decreases rapidly with an increased number of complementary sex-determining loci.*

use such opportunities to take advantage of a superior system with a higher fitness as other hymenopteran species have done? The answer may be that the colony gets by, but the price it pays seems very high indeed given the extreme polyandry of the queen that is probably a direct evolutionary response to the genetic load associated with the *csd* locus and buffers the negative effects of diploid drones at the colony level (Page 1984). Even the cannibalism by workers seems to be a derived behavioral trait in response to the single-locus system because diploid drones are known to reach adulthood in other bee species. So clearly we are left with a puzzle, but it is a smaller part of the larger puzzle of the evolution of sex.

8.4. The lazy drone paradox

Drones are supposedly lazy and do not engage in any contribution to the work of the colony (Figure 8.4). This is a phenomenon observed quite often in social hymenoptera, where the females do the work and the only purpose of the males seems to be mating. Indeed, the males are a special case. If they could go to pubs, we might expect to see them there as well. They can collect neither pollen nor nectar and have no sting and no wax glands. So the bauplan message is very clear: We can neither work, nor build, nor fight and can only mate. But why should the bauplan be like that? Why not have males that contribute to the workforce? In termites, male workers happily participate in all activities required to keep the colony going. Clearly, the drone's genome contains all those most efficient genes that would allow them to work if only they were expressed in an appropriate way. Why has evolution been so cruel as to deny them these

FIGURE 8.4 *A lazy drone?*

opportunities? They are about twice as big as workers, and it would seem so adaptive if they could only do a little bit of work.

8.4.1. SIZE MATTERS

Evolutionary pressures seem to have fiercely suppressed any characteristics expressed in workers, in the male sex. Indeed, the male's bauplan seems to be an optimization within the honey bee genome aimed at three main objectives: finding receptive queens, flight performance, and sperm number. Drones need to carry as large a sperm payload as they can, as far as they can. Indeed, drones are essential for establishing gene flow in all colonial insects studied to date. Generally, the males are the strong fliers, the queens are the strong breeders (Paxton 2005), and the workers do the rest. Males need only be good with regard to mating success, nothing else.

When the males are reared, the colony is at its seasonal peak, preparing for swarming, and hence at a time when worker numbers in the colony are high. Additional worker brood is not needed to further prepare the colony for swarming. So the males should not work but, rather, spread their genes and only mate. Feminists may consider this to be a dark plan, and they are correct. Males are costly, particularly because size matters for them. Bigger drones have been shown to have more semen, conduct longer mating flights, and have a higher mating success (Berg et al. 1997; Schlüns et al. 2003, 2004; Couvillon et al. 2010). So the question is not so much why drones do not work but, rather, why are they not bigger to be even better and have more semen? Indeed, males of *Dorylus* army ants can exceed the size of workers by almost an order of magnitude (Figure 8.5). The workers do not fly but forage underground, whereas the males are strong flyers that are essential for maintaining the gene flow between populations over long distances (Barth et al. 2013). Honey bee workers do not seem to care very much about size when interacting with drones (Goins and Schneider 2013). However, workers themselves control the size of the drones by constructing the large drone cells in the comb. These are almost 40% wider in diameter than the worker cells (Hepburn 1986). Drones can also be reared in worker cells; however, then they are smaller than regular drones. Drones larger than normal might be possible if there were only bigger cells. Sometimes drones occur in queen cells because of a mislaid haploid egg, but

FIGURE 8.5 *Size dimorphism between a Dorylus worker (bottom) and a male (top).*

then these drones do not develop into adults since the developmental pathways are disrupted and their development is arrested in the late larval stages. In fact, we do not know how large a drone can potentially become. Clearly, there must be some trade-off at the colony level between the number of drones produced (it should be many) and the size (they should be fit) because there is only one comb on which a certain number of cells can be modified for drone production. So it will be either a few large ones or many small ones. The number of sperm alone seems insufficient to address this issue. The reduction in sperm number of drones reared in small cells is not compensated for by the increased sperm load in the larger individuals.

Of course, there is more than just sperm number when it comes to individual fitness of a drone. Flight performance may easily be more important for mating success than the number of sperm. Colony fitness is about producing successful males because the probability for mating is so minute and the competition so high. We view this as a most interesting topic for study, determining the optimal adaptation of drone cell size and testing whether the bees get this correct on most occasions or whether they are faced with another set of compromises.

The male bauplan confirms that it is only mating that they are designed for. Their odor receptors are specialized to detect a single molecule of the queen's pheromone (Brockmann, Brückner, and Crewe 1998; Wanner et al. 2007), and their antennae have 10 times the number of antennal plate organs as found in workers (Esslen and Kaissling 1976). The much larger compound eyes comprise twice the number of facets found in workers (Seidl 1982). The upper part of their eyes is specifically designed to spot the queen in flight with regions of exceptional high spatial resolution. Both temporal and spatial resolution are twice as high as in workers (Menzel, Wunderer, and Stavenga 1991; Streinzer et al. 2013). Their flight muscles are twice the size of those of workers, suggesting that the drone's flight apparatus is made to facilitate the special requirements of a successful mating flight. Indeed, they carry a precious payload. The endophallus is not only among the most elaborate of male mating organs (Figure 8.6) but also part of an exceptionally massive reproductive tract weighing 27.8 mg, comprising 13% of the total body mass.

FIGURE 8.6 *Endophallus of a honey bee drone. The semen is at the top of the structure.*

8.4.2. FLIGHTS OF FANCY

Drones carry on average 16.1 mg of honey when taking off for a mating flight (Coelho 1996). This is equivalent to an average nectar payload of 44 mg, which exceeds that of a pollen or nectar forager (Winston 1987), but only by a marginal 10%. Compared to worker bees, drones are not very fuel efficient. Environmentalists would consider their carbon dioxide (CO_2) footprint as an ecological disaster. To fly at their average speed (19.3 km/h), they burn approximately 1.24 mg of honey per kilometer, resulting in a CO_2 footprint of 1.5 mg of CO_2/km. This does not compare favorably with the workers' efficiency: They fly at an average speed of 21.9 km/h but can achieve this while producing 0.63 mg/km or 0.75 mg CO_2/km carrying the same payload. Compare this to a 1.5-ton sedan automobile (7 l/100 km, ~6.5 million drones with a payload of 290 kg). It produces approximately 2.7 kg of CO_2 per liter petrol, leaving a footprint of approximately 1.61 kg CO_2/km. The weight equivalent of drones operates at approximately 9.5 kg CO_2/km, which is in the range of a modern truck that consumes approximately 25 l/100 km in short-haul regional traffic with a substantially higher payload (Martensson 2010). The workers with the same payload are in the range of the automobiles of the 1970s, running at approximately 12 l/100 km. So the larger size does not seem to be important for carrying a larger payload. A worker-sized drone would very well be able to transport the extra weight of the genital tract at a higher speed a further distance using less energy. So is the larger size of the drone a maladaptation? Considering the maximum flight speed, it seems so. Drones and workers have a rather similar maximum flight speed of 28.8.km/h. It seems clear that the drone's flight ability has not been designed by evolution to conduct long-distance or long-duration flights, as the worker's bauplan allows for much longer flights. The average drone flight range has been estimated at about 6 km, six times less than that of a worker (Coelho 1996). So the large size of the drone is neither about long flights nor about fast flights. Oh these wasteful males!

However, all is not lost. Another major difference between workers and drones in flight characteristics is the thrust (F_{max}) of the flight muscles that relates to the maximum lifting capacity. The thrust is more than twice as high in drones (3.95 mN) as in workers (1.73 mN) and higher than in queens (2.15 mN) (Coelho 1991). The lifting force is directly related to size, and drones can lift up to 145.8 mg weight in flight (Betts 1939, as cited in Coelho 1996) in addition to their reproductive tract. This is a huge payload of more than 50% of their own body weight to carry in flight, which exceeds the maximum 30% payload of a standard Chinook helicopter by far. They are much stronger flyers than is needed to be able to just carry their reproductive tract. For semen transport alone, worker-sized drones would be entirely sufficient. Why are they so strong? It seems safe to discard the notion that they might appear beautiful to the queen as a result of cryptic sexual selection. The males search for the queen and approach her from behind for mating, providing little, if any, opportunity for some beauty (or other) competition as occurs in higher vertebrates. The likely explanation has to be sought in the mating itself, which may require extraordinary capacity for lift. Mating takes place in flight, and the coupling of drone and queen is tightly linked until the endophallus disrupts and the drone falls down. In order for the couple to stay in air, this will indeed require rather extraordinary flight characteristics. The queen is clearly too weak a flyer to achieve this (Coelho 1991). This becomes clear whenever the drone fails to separate after mating. The queen then falls down to the ground together with the dead male (Gary 1963). It may be the strong drone with a lifting capacity of 150 mg, close to the weight of the queen, that keeps the couple aloft. The idea of the strong drone holding the queen is rather old indeed and was put forward by the most famous evolutionary biologist, Charles Darwin. Although Darwin (1890) did not know much about the details of the mating biology of the honey bee, for him it seemed clear that "the male requires great strength and size in order to carry the female through the air" (p. 279). This may have appealed to the male chauvinism of the day, but it did lead Darwin to a rather plausible conclusion.

8.4.3. DEATH IN MATING

The drones are truly heroic. Their main purpose is mating with the queen in flight. They die when inflating their endophallus with almost all the hemolymph they have. So they immediately die of shock and can only mate once. This seems to be another very grim and dark side of honey bees—this time the male's life seems rather unnecessarily wasted. Why not stay alive to mate with yet another queen or two? Even a little bit of semen in another queen would seem so adaptive. However, given the extremely low probability of finding a queen to mate with in the first place, it may well be adaptive for a drone to expend all (including his life) rather than retain sperm for a second mating that will never happen. The numerical sex ratio is extreme in honey bees. Some few thousand drones are produced in a season but typically only one or two queens per colony. So the average drone suffers a much worse fate than those that die during mating: They die without mating. Although this may sound sad and tragic for the individual drone, it is the rule rather than the exception for the haploid males of most higher biological systems that have sexual reproduction. By definition, the female gametes are the macrogametes and the male ones the microgametes. Typically, the latter are produced in excessively large numbers. Just think of the life of an individual sperm. Almost all of them are wasted and only a few can fertilize an egg. Here, it pays to consider the colony level and address the swarm with the virgin queen as the female unit and the drones as the sperms of the colony. The colony's investment in females is massive and comprises most of the individuals in the colony. The males are only produced when there are excess resources for colony growth and the colony does not compromise its growth by rearing drones. Although the cost of rearing drones is often overestimated by beekeepers, rearing drones is clearly not completely without cost to the colony.

8.4.4. PARASITIC DRONES

In order to achieve sexual maturity, the drones need to be fed and cared for by the workers through all their stages of development including as adults. They receive proteinaceous food jelly from the hypopharyngeal glands but also honey and pollen from the workers (Free 1957; Crailsheim 1992, 1998) so that they weigh twice as much as the workers and even more than the queen (Es'kov and Es'kova 2013). If there are many drones, many workers are needed to take care of them, which inevitably constrains workers that then cannot engage in other tasks that might also be to the benefit of the colony or simply might distract them from their well-earned working break. The colony therefore very carefully regulates the number of drones produced (Wharton et al. 2007). As soon as conditions turn negative, the workers immediately stop rearing drone brood and cannibalize it instead. Apparently, drone production is viewed as a luxury that can only be indulged when the colony is doing well with sufficient excess resources.

It is here that colonies of some honey bee subspecies engage in a most efficient but arguably dirty trick. They outsource the feeding of and the care of drones to other colonies. This happens through the drifting of adult drones between colonies. After an orientation or mating flight, the drone does not return to its home colony but, rather, to a different one. In European honey bees, this is thought to be simply due to orientation errors in a process very similar to the drifting of workers and queens. As long as all colonies do this in a similar way, then this does not result in a major fitness advantage for specific colonies. However, if drones of some colonies start actively searching for foreign colonies and others do not, then this will clearly result in a fitness benefit for the parasitic drone strategy. This has been studied in detail by Thomas Rinderer and coworkers (1985) in the context of the spread of the Africanized honey bee in the

Americas. Whereas drones of European honey bee colonies remained in their home nests, drones of the Africanized honey bee colonies were found much more frequently in European colonies. So apparently the African colonies were more efficient in preventing foreign drones from entering their colonies and also sent out more parasitic drones to European colonies. This reduced drone production in the new host colonies while allowing for more drone production in their own colonies.

Nevertheless, it seems that there is relatively little hybridization between European and African honey bees in the Americas because the African bees seem to spread as pure African lineages (Pinto et al. 2005; Rangel et al. 2016). This may be because they are adept at survival in the wild in tropical and subtropical climates and produce parasitic swarms that take over colonies. On the other hand, European honey bee colonies are largely restricted to the apiaries of beekeepers and do not show this behavior.

8.5. The multiple mating controversy

The queen mates with many males—in most European subspecies between 10 and 20 drones and with larger numbers in other subspecies. This seems odd because it reduces intracolonial relatedness, and a high relatedness is one of the main arguments for the evolution of sociality. Yet clearly hymenoptera became social when females were mating with single males, and polyandry only evolved after sociality had been established. It is a derived trait typical of social hymenoptera with large colonies (Cole and Wierznas 1999). The debate about honey bee queen polyandry dates back to the days of Swammerdam and de Reaumur. Arthur Dobbs (1750) was the first to champion the sperm limitation hypothesis: "It seems therefore necessary that the Queen should breed for many Males, as by one Act of Coition from each, may impregnate all her Eggs." He correctly interpreted the spermatheca (labeled *g* in figure) to hold the sperm load of the queen in de Reaumur's (1750) Plate 32, Figure 5 (p. 520), which had been drawn after Swammerdam's Table 19, Figure 3; Swammerdam was quite precise about "the eggs are smeared over before the Bee lays them" (Swammerdam 1737, 1758).

As is often the case in science, it took a while to revive the issue. It was only after William "Bill" Hamilton (1964) revolutionized the theory of social evolution by developing the inclusive fitness theory that students of social insects regained an interest in the adaptive value of multiple mating. Whereas the haplodiploid population structure allowed for extremely closely related super-sisters ($G = 3/4$) that should greatly facilitate the evolution of altruism, the honey bee queen seemed to ignore this brilliant theory by mating with an endless series of males, drastically reducing the average intracolonial relatedness close to $G = 1/4$. So clearly any benefit of polyandry had to lie at the level of the colony. During approximately the past 20 years, there have been a large number of plausible explanations as to why the queen mates with so many drones, resulting in a lively debate. These range from Arthur Dobbs' explanation of a simple need for sperm in large colonies to that of the benefits of intracolonial genetic variance that contains rare subfamilies that might specialize in rare but important tasks. We do not want to engage in a detailed comparative analysis, and interested readers are referred to comprehensive reviews in this regard (Brown and Schmid-Hempel 2003; Oldroyd and Fewell 2007; Jeanson and Weidenmüller 2014). We nevertheless briefly introduce readers who do not have the time, energy, or motivation to familiarize themselves with the entire debate to the essential critique against the arguments. Furthermore, the empirical evidence for the positive impact of polyandry and genetic diversity is not unequivocal. Although an increasing number of studies have found a positive effect of multiple mating on colony fitness (Fuchs and Schade 1994; Mattila and Seeley 2007), others have not (Underwood, Lewis, and Hare 2004; Bratkowski et al. 2012), suggesting that the effect may not be very strong or variable or both (Palmer and Oldroyd 2000). There is also an intrinsic academic sociological issue resulting in a potential bias in the publication of research. Those studies that produced results that showed no significant effects of the queen mating frequency on the colony phenotype may not only have failed to pass peer review but also not even have been written up by the scientists and potential authors who got frustrated with their negative results and decided to consign them to their bottom drawers.

8.5.1. ARGUMENTS AGAINST THE GENETIC VARIANCE HYPOTHESES

Hypothesis: The more genetically variable the members of a colony, the better.

Honey bee queens hold the world record for multiple mating, as far as we know. The record was published for queens of the Asian giant honey bee *Apis dorsata* by Wattanachaiyingcharoen et al. (2003), who reported the queens had up to 102 matings. Clearly, this reduces the intracolonial relatedness to $G = {\sim}0.25$. But already at ten matings, the intracolonial relatedness is as low as $G = 0.27$. Little is gained by the extra 90 matings in relation to intracolonial genetic variance. In a comparative analysis, Jaffé et al. (2010) found that species with highly polyandrous queens, such as the honey bee, had a low paternity skew, with paternity equalized among potential sires. An even representation of the males in the offspring of the queen is expected to maximize fitness benefits derived from genetic diversity among offspring and would indeed be predicted by the genetic variance hypothesis. However, we disagree that this can resolve the issue at this stage. A low paternity skew is interesting but does not address the previously presented critique. The fundamental question is whether taking the additional mating risk is compensated for by a higher gain in fitness terms. The mating risk is most likely a stepwise function (Fuchs and Moritz 1999) because every mating flight poses a high risk, whereas the individual matings during a mating flight that occur within minutes in short succession may be less critical. In many models, the cost of multiple matings is either not considered or assumed to be negligible. We believe this is a weakness in the theory.

8.5.2. ARGUMENTS AGAINST THE SPERM LIMITATION HYPOTHESIS

Hypothesis: The queen needs a lot of sperm to sustain the colony.

The strongest argument against the sperm limitation hypothesis is that a single drone provides sufficient sperm to fully inseminate a queen. Indeed, the queen expels most of the semen from individual drones after mating, and only 4–8% reaches the spermatheca. This seems not particularly adaptive under the sperm limitation concept. So the question is whether the queen is just hopelessly inefficient or whether she does this on purpose in order to achieve a diverse mix of sperm. The interpretation is not easy, particularly because evolution is known to solve problems pragmatically but not necessarily in the best way. A functional way is evolutionary sufficient if the species gets by. A change in the bauplan of the queen to handle sperm more efficiently may come at a considerable cost. There may be trade-offs with regard to flight ability and body size. Hence, evolution may not be in a position to modify the bauplan to achieve an optimal solution with regard to sperm transfer. So although queens of various stingless bee species that have much larger colonies than the honey bee are known to be singly mated, this does not help us in the argument regarding the honey bee. As we have previously shown, stingless bees are also much more efficient in the swarming process. Now we see that they are also much more efficient in transferring sperm. So are they better bees because they have colonies with much reduced intracolonial diversity and often perfect Hamiltonian intracolonial relatedness? Unfortunately, the argument remains inconclusive. One can obviously argue with great plausibility that because the very large colonies of *Scaptotrigona postica* function superbly well, genetic variance is obviously completely irrelevant for enhancing colony fitness. So we are easily stuck if we employ comparative analyses. Multiple mating has evolved several times independently, and it may very well be that this occurred for very different reasons and specific to the species in question.

Very often, scientific authors prefer to present only one side of the coin in this academic discourse favoring the one or the other side. It seems we (the authors of this book) will probably see the debate continue for quite some time without a clear-cut answer. However, this may be exactly what biology is about: Several solutions can be found for the same problem, and none of them may be an evolutionarily stable strategy in the strict definition of John Maynard Smith (Maynard-Smith and Price 1973; Maynard Smith 1981). After all, evolution is a continuously dynamic process. We clearly see only a minute finite temporal window, can gather some evidence to speculate about the past, but certainly would not dare to predict what honey bees will look like after another million years of natural selection and evolution.

8.6. Conflict in queen rearing

Irrespective of the reasons for the evolution of multiple mating, one thing is clear: It will dramatically enhance the potential for intracolonial conflict. This is most striking during queen rearing because the subfamily providing the new queen will win both direct and indirect fitness, whereas the others must settle for the indirect part. So who chooses the new queen? Obviously, the nurse workers decide who is to become a queen and who is not by the feeding regime supplied to the royal larvae (see Chapter 5). Yet there are two very different occasions when the colony embarks on queen rearing that are regulated in fundamentally different ways. One is during reproductive swarming, during which the old queen leaves and new ones need to be reared, and the other is during so-called emergency queen rearing. The latter happens when the old queen is accidentally lost (e.g., by the careless beekeeper's apprentice, who squished it when removing a frame). But it could also be that the queen dies from some other unfortunate event. Planned swarming and conditions of unforeseen queen loss are very different from each other.

During swarming, the old queen leaves the colony in a prepared manner. About a week before the swarm leaves the colony, the workers build specific cells for the queen to deposit eggs into. The larvae from these eggs will then receive the royal diet and develop into new queens. Hence, although the workers are ultimately in control, the queen deposits eggs into a very limited number of prepared queen cells. In European honey bees, some dozen of queen cells may be prepared. In this case, the queen can decide which egg to lay into one of these queen cells. Workers will not remove an egg or a larva from a queen cell unless it is dead or diseased. This puts the queen in a decisive position. In theory, workers might have the opportunity for nepotistic queen rearing, but instead they take what the queen provides, which is an egg fertilized by a sperm that is a random sample from the multiple drones she mated.

In emergency queen rearing, this is not the case. Now the workers can choose from a very large number of larvae that just need to be at the right age such that they can still be reared as queens. Indeed, it seems this results in a very different solution, albeit not in support of inclusive fitness theory. If the workers tried to maximize their inclusive fitness, there might be havoc in the colony, with all workers removing the eggs of the other subfamilies. This obviously does not happen because workers are unable to discriminate between super-sisters and half-sisters, as noted in Chapter 6. Hence, one might assume that in emergency queen rearing the most frequent subfamily in the colony will provide the next queen. However, this is absolutely not the case, and in a series of experiments it was shown that exactly the opposite is true: Larvae from extremely rare subfamilies are preferentially chosen to become the next queen (Moritz et al. 2005; Lattorff and Moritz 2016). So there is a huge difference in the nature of queens reared in the swarming process compared with emergency queen rearing, and the workers are clearly not conforming to inclusive fitness theory.

So what could be a plausible explanation for such a seemingly completely maladaptive behavior? Why should workers rear the larvae of the rarest subfamilies if they are only related to them by 0.25 instead of 0.75? Obviously, mechanisms other than relatedness or random chance are important drivers. However, we should not be surprised. Differential treatment of larvae is well known in the honey bee colony. This can be with respect to sex or health status but also to reproductive potential. Beekman, Calis, and Boot (2000) reported on preferential rearing of parasitic *Apis mellifera capensis* worker larvae introduced into European honey bee colonies. The workers fed these larvae significantly more, even if they were reared in worker cells. Indeed, this food regime may be an important mechanism that allows these workers to develop into more queen-like individuals as adult bees (see Chapter 6). So if the nurse bees in emergency queen rearing situations choose those larvae that are particularly attractive to feed, this might explain why larvae of specific subfamilies are preferentially reared. However, this leaves us with the question of why these royal families are not more frequent in the population if they are selected during every emergency rearing event. A plausible explanation may lie at the colony level of selection, as we have already demonstrated in the Chapter 6. Once the royals are frequent, the workers become less frequent. So if colonies headed by these rare royal family queens produce less efficient workers, they might produce fewer sexuals, fewer swarms, and may thus have a reduced fitness at the colony level. In 1989, Hillesheim, Koeniger, and Moritz demonstrated the trade-off between individual fitness and colony failure. Colonies composed of reproductive workers failed to engage in brood care and hence collapsed, which most plausibly explained the *capensis* calamity in South African apiculture more than a decade later. An alternative trade-off might be that emergency queens produce less fit males; thus, the fitness benefit of the one sex is offset by the fitness decline in the other. This has been reported for *Drosophila* fruit flies (Chippindale, Gibson, and Rice 2001). Currently, there is no empirical evidence for a mechanism of this kind in honey bees, but the reproductive fate of the emergency queens vs the swarming queens needs to be explored in greater detail.

8.7. The drone congregation area

Queens and males mate at so-called drone congregation areas (DCAs). Because drones fly out of the hive well before the queens do so, the classical understanding is that the drones gather at these DCAs well before the queens arrive. Indeed, this seems to be highly adaptive at the colony level, which produces males in excess but only very few and therefore highly precious queens. Out in the field, drone aggregations are often acoustically identified by the specific buzz that is in the air when hundreds of drones circle at the site. Throwing a small stone in the air can then release the typical comet of hundreds of drones pursuing the object in an attempt to achieve mating. So once at the DCA, obviously optical cues are important for the drones to pursue the flying queen. In addition, queen pheromone lures can be used to attract the drones. Drones follow small, black queen-sized objects that are coated with the queen's mandibular gland pheromones in a frenzied fashion.

This behavior can also be used to catch drones in a trap net with several pheromone lures inside it (Figure 8.7). The trap is lifted into the air by a helium balloon or a quadcopter to approximately 15 m above the ground and maintained in this position until a sufficient number of drones have been caught. The drones fly into the net from below and aggregate at the top, unable to fly down to exit the trap. This provides an elegant means of obtaining population-wide samples because the drones from many different colonies aggregate at these sites. The locations of the DCAs have been shown to be stable over time, which is most useful for studying the dynamics of the population composition from one year to the next. The drones embark on their mating flights well before the queen; hence, all the previously mentioned evidence seems to suggest that the drones first fly to the DCA in expectation of the subsequent arrival of queens. In this scenario, the males are establishing a lek (i.e., DCA) to which the females are attracted. However, we believe that

FIGURE 8.7 *Field assistant attempting to catch drones using the William's trap attached to a helium balloon and held with a "decapitated" fishing rod.*

the conventional wisdom of many textbooks and monographs that would want us to believe this scenario may actually be incorrect, and we list a few arguments that the situation may in fact be the reverse—the queen attracts the males, and the males are not waiting for her to come to their DCA. Rather than the drones establishing a lek, we believe the males are attracted to queens due to the female calling syndrome (Hölldobler and Bartz 1985; Hölldobler and Haskins 1977).

The main problem with all DCA studies is that they cannot rigorously exclude the possibility that queens from unknown colonies had already been at the DCA before finding the site. Just screening flight time at the flight entrance of a colony is insufficient to conclude that the drones might end up at a common DCA. They may fly out in the same general direction depending on local conditions, including wind, landmarks, and flight experience (Loper, Wolf, and Taylor 1992), but this does not indicate where they start circling to wait for the queen.

DCAs have been defined as sites where the drones can be attracted to a queen or a pheromone lure. The boundaries are defined whenever drones leave the lure and can no longer be attracted. This is again not conclusive because it is not possible to test this with a few lures/queens only. Drones will eventually stop pursuing a lure when matings cannot be performed. After a few unsuccessful mating attempts, which are energetically highly demanding, drones will of course return to the hive and no longer follow the lure. Because drones follow the queen by optical and pheromonal cues, any queen in flight arriving at the site will also distract the drones from the lure, again making the loss of drones at a certain "boundary" inconclusive. A dense network of lures on the DCA might be required to assess this problem in an unbiased manner. We know that drones are neither randomly nor evenly distributed at a DCA. Given the exceptional small chance of mating success, they should pursue a virgin queen at any point of a mating flight and not wait until a queen has reached a specific DCA. However, if queens can control the release of their sex pheromone until they are in the vicinity of the DCA, then the drones will be confined to pursuit of mates in this area.

Most important, if it is the male sex that attracts the female, the males typically develop very strong optical, acoustical, or chemical signals to attract the female (not only in insects). This is a result of very strong sexual selection. To date, such signals have not been identified in drone honey bees. So what magic should the queen use to find the DCA with the eagerly waiting drones? Many studies have failed to provide unambiguous answers. On the contrary, we do know that the queens produce a highly efficient pheromone to attract the males. Males have highly sensitive and highly specific receptors to receive as little as a single molecule of 9-oxodecenoic acid (9-ODA), the major compound of the queen pheromone. It is the males that are able to follow the queen because of their excellent compound eyes and flight ability. So from a theoretical perspective, it seems the queen is in the better position to attract the males rather than the other way around. Indeed, we conducted some pilot studies to address this issue by monitoring wind direction and an array of drone traps spread out on a known DCA site in the Ezemvelo Nature Reserve in Gauteng, South Africa. We were able to move the DCA upwind but not downwind, suggesting that indeed the spread of the pheromone is essential to attract the drones. The males flew along the pheromone gradient against the wind and were always caught in the first drone trap encountered and not at the same site as the day before. Similar observations were made by Gary Loper and co-workers (Loper, Wolf, and Taylor 1993), who used radar technology to spot the drones. A queen pheromone source caused a line orientation of the flying drones downwind of the pheromone source and attracted the drones from almost 1 km away.

So where is the queen and how does she know where to fly? Again, experience obtained during orientation flights and local landmarks may be important. However, if we assume the queen wants to spread her pheromone as far as possible, she should fly on the way out with the wind. That way, the pheromone plume would be carried farther and she might attract more drones. The drones, on the other hand, should fly upwind, perhaps circling, to facilitate the sensing of a pheromone signal even from distant queens flying downwind toward them. This mechanism would also be most efficient to avoid brother–sister matings because queens and drones from the same colony would depart in different directions, local landscape constraints permitting. We do not know if this hypothesis applies to the honey bee mating system, but clearly it is not in conflict with any current empirical evidence on honey bee mating flights. Given the current paradox that the female has a sex pheromone but the males are supposed to wait for her at some predefined place, we believe it is time to critically re-examine mating flight behavior. Radio-frequency identification and harmonic radar technology can be used to accurately account for flight activity and trace the flight path of drones and queens over several kilometers.

8.8. The queen's mating flight

Another major maladaptation, however, seems to lurk in the mating flight of the queen. In large monogynous perennial colonies, the queen is of exceptional value and natural selection should favor traits that put the precious queen at as little risk as possible. Again, stingless bees seem to have solved this problem in an elegant fashion. The virgin queens stay in the colony, and only the males embark on mating flights. The males assemble in tight aggregations in front of colonies awaiting virgin queens to come out. These male assemblies disappear overnight only to reform the next day (Figure 8.8). They are composed of males unrelated to the colony where they aggregate, hence actively avoiding inbreeding (Lopez and Kraus 2009). So why should the honey bee queen not follow a similar procedure—wait for the males

FIGURE 8.8 *Male congregation of Scaptotrigona mexicana in front of a hive on a meliponario in Tapachula, Mexico. The congregations disappear at night to reassemble the next morning.*

to aggregate in front of the colony, rush out, get mated, and return to the colony? What is the advantage of flying several kilometers in elegant circles in order to attract drones if drones of other species are much more polite and greet the queen right on the front doorstep? Wouldn't it also be highly adaptive for a honey bee drone to just wait there for the queen to come out and then intercept her to make sure he is the first male mating? Sexual selection on both sexes should try to minimize the risks of mating. It may again be the bauplan constraints that the honey bees have not been able to escape, whereas the stingless bees have done so.

Clearly, the honey bees evolved from species with an annual colony life form with a solitary overwintering queen founding a nest in spring (haplometrosis), probably similar to bumblebee species. Hence, mating took place in the previous year with a single male before the queen searched for a suitable place to hibernate. In such a life history, the risk of mating is minute compared to losing the queen over winter or failure to establish a colony in the next year. Nevertheless, males need to find females, and a sex attractant might greatly facilitate mating success (female calling syndrome). So a pheromone is in place, as is a mating flight, and both are fixed elements of the life history cycle. To get rid of these traits because a queen becomes much more important may not always be easy. Evolution builds upon existing characteristics and does not start from scratch (which is why it is not called revolution). Hence, old patterns will be modified until they suit a new life form to have reproductive success. So if a queen has a highly specific sex pheromone with low volatility, she may not attract very many males if she stays inside a sheltered nest cavity. We do not know if stingless bee colonies chemically advertise the presence of virgin queens to the males, but clearly there needs to be some signal getting to the outside of the colony. The drones do not enter the colony for mating but, rather, patiently remain next to the flight entrance outside the nest. Honey bees opted for a more risky solution to deal with the ancestral constraints, and we are not convinced it is a very elegant solution. Particularly in light of the extreme degree of polyandry that often requires multiple mating flights by the queen, it seems like an unnecessary risk that could easily be minimized if *Apis* had thought of a solution similar to that of stingless bees or several ant species with large colonies in which the queens do not take risky mating flights on their own but, rather, always stay in or in close vicinity of the nest.

If honey bee mating behavior has been jury-rigged around the female calling syndrome, then the nature of its mating behavior needs to be carefully re-examined using the latest technology.

{ 9 }

Apiculture and Long-Suffering Bees

9.1. Search for the "perfect" bee

Apiculture is several thousand years old and yet honey bees have never been "tamed" like pets or other domesticated livestock. There may be many reasons for this, including the complex mating system that makes controlled mating an extraordinary challenge (Koeniger et al. 2014; see also Chapter 8, this volume). A major problem for many breeding programs is that obtaining an adequate level of control over the males participating in mating has been difficult to achieve (if attempted at all). Selection has therefore focused on the female lineage, which is easy to control, with the males remaining completely uncontrolled and unknown (Oxley and Oldroyd 2010). The queen's promiscuous mating with many males (Estoup, Solignac, and Cornuet 1994; Kraus, Neumann, and Moritz 2005) has serious consequences for colony functioning, and the variety of genetic contributions by the males to a colony has been shown to be important for colony-level phenotypes (Neumann and Moritz 2000), including disease resistance (Seeley and Tarpy 2007). Strong genotype–environment interactions determine colony efficiency (Costa, Lodesani, and Bienefeld 2012). This makes the progress of selection in breeding programs both difficult and slow but not impossible, as shown by several successful breeding programs for specific strains (Page 2013). Indeed, honey bee breeding has contributed to the production of various commercially successful breeding lines.

Honey bee breeding often pursues various goals at the same time. The key attributes sought by the breeders are honey production (often the foremost goal), but also important are behavioral traits, including a low propensity for swarming and defensive behavior. Ideally, beekeepers want honey bee colonies that do not swarm, produce a lot of honey, but do not sting. As a consequence, they are in search of the holy grail of the best bee that exhibits all these traits. The most extreme form of this quest and at the same time, from the perspective of selective breeding, least viable concept was that of Brother Adam of Buckfast Abbey (Adam 1983; Figure 9.1). Brother Adam said that he

FIGURE 9.1 *Karl Kehrle, alias "Brother Adam" (1898–1996), the creator of the "Buckfast bee."*

> was looking for nothing other than the "perfect bee," that is, one particular race of bees which surpassed all the others in a combination of characteristics of economic value, especially for honey production. Such a search would of course have been a fruitless venture, for Nature never breeds for the perfection of factors we desire for our commercial needs (p. 11).

He (Adam 1983) was well aware that natural selection was unlikely to help him in this endeavor because "Nature's aim is almost exclusively for the preservation and multiplication of a type" (p. 11) and "Nature with the means at her disposal has in no way produced a 'best bee' or an 'ideal bee,' still less a race of bees which answers all the desires and needs of the modern beekeeper" (p. 206). He stated that

> the synthesisation of new combinations through cross-breeding is, indeed, the only breeding worthy of the name. It alone allows us to actualise all of the potentialities involved. For it alone has the power to bring together all the various races and strains with their desirable economic qualities, combining these into new types of bee, while at the same time eliminating the deleterious traits, and thus producing a bee that will completely answer all the needs of modern beekeeping. (p. 206)

However, is this a viable approach? Is it a valid philosophy to propose a global search for the best strain(s) of bees? We think not. We consider this approach to be fundamentally flawed, and it may harbor one of the major misconceptions of honey bee breeding. This is not about the dark side of the hive but, rather, about a less enlightened side of honey bee breeding. In Brother

Adam's case, he labored against the reality of mating biology because any queen of a "perfect bee" would need to mate with regionally available imperfect males, frustrating most breeding efforts in just one generation. Bee breeders have been aware of this problem and used mating apiaries and offshore islands where they place virgin queens and drones, providing colonies for controlled matings. However, even on the islands, matings with mainland drones can occur (Neumann et al. 1999), interfering with successful breeding. Alas, the main problem is that the number of queens that can be produced on islands is too small to saturate entire populations on the mainland. In addition, there are problems resulting from applying classical concepts of animal breeding to an organism that is simply not suitable for standard breeding procedures. The colony is a genetically heterogeneous entity comprising a complex combination of genotypes, and we can only reliably control the male side using complicated instrumental insemination requiring highly skilled technicians (Cobey and Schley 2002). But even given that this technique could be implemented at a sufficiently large scale to establish homogeneous apicultural populations successfully, is the goal of producing big and gentle colonies a sustainable concept suited to both apiculture and nature? Today, we know that pollination is the most important service provided by honey bees for both the agricultural economy and ecology (Aizen and Harder 2009; Gallai et al. 2009); thus, it seems anachronistic that the prime selective breeding trait should be honey production. Although honey production is important for most commercial beekeepers, we must keep in mind that the economic value of pollination services provided by honey bees far exceeds that of the direct income of the beekeeper selling his or her bee products. The benefits to beekeepers, agricultural industries, and human society are important but pale into insignificance compared to the requirement for ecosystem services that need honey bees that excel in pollination rather than in honey production. Fortunately, beekeepers aim for large colonies that are good at both pollination and nectar foraging. It is primarily the sheer numbers of bees that matter to achieve successful pollination. A pollinating honeybee is not a lonely forager.

9.2. The fallacy of the big, gentle colony

So why are we concerned that selection for size, docility, and honey may cause trouble and may not be sustainable? The price to be paid may actually be too high to allow for sustainable beekeeping. A major outcome of selection is missing in classical breeding schemes: selection for disease resistance (Moritz et al. 2010). Strong colonies that do not swarm and are not defensive are favored not only by beekeepers but also by pests and pathogens. Low nest defense allows pests to enter the colony more easily, and selection for docile colonies may facilitate this effect. Active swarm prevention results in large, long-lived colonies with big brood nests—ideal for the spread of any brood parasite and pathogen. Beekeepers try to overcome the problem of creating these ideal pathogenic conditions by applying treatments to fight the diseases. Although this helps the colony in the short term, it relaxes any natural selection for potentially resistant honey bees. In times of concern about the global impact of colony losses, honey production per colony should actually become a secondary selection goal, and colony health and survival must become the primary objective. Indeed, increasing attempts to engage in breeding resistant bees have shown promising results (Büchler, Berg, and Le Conte 2010). A particularly instructive example is that on the island of Gotland, where in 1999 Ingemar Fries established a population of 150 colonies infested with the brood parasitic mite *Varroa destructor* and then did nothing but wait (Fries et al. 2003; Fries, Imdorf, Rosenkranz 2006; Fries and Bommarco 2007; Locke and Fries 2011). After only seven years had elapsed and after having lost most of his original population, he had established a strain of bees with resistant traits by just letting natural selection operate. Indeed, natural selection has shaped a huge variety of subspecies and ecotypes (without Ingemar Fries interfering) that show highly specific adaptations to regional environmental conditions. Brother Adam's quixotic quest for the "best" strain of bees was based on extensive sampling of this genetic diversity without appreciating that this diversity needs to be conserved as the outcome of local selection for environmental adaptation. Local conditions impose divergent constraints on the life history of a honey bee colony throughout a year. Honey bees in the tropics have very different adaptations from those in a temperate climate. Habitats with highly dynamic, rapidly changing food resources will require completely different strategies from those with highly predictable conditions. From classical ecological theory, we know that the first condition (variable short-lived environment) will drive species to invest more in reproduction, whereas species in stable environments should invest more in growth rather than reproduction. The various honey bee subspecies seem to exemplify these predictions perfectly, as revealed by the comprehensive work of Friedrich Ruttner (1988) and discussed in the following sections.

9.3. European honey bees

The biogeography of *Apis mellifera* is well understood (Figure 9.2). Honey bees in Europe are exposed to extremely variable but also highly predictable food source availability throughout the seasons. Winters are poor but summers are rich. The more northern subspecies are exposed to particularly long winters and short summer seasons, when forage is massively available but only during a narrow time window. To take full advantage of these seasonal dynamics, colonies need to be strong early in the season. This in turn means that they need to be strong before entering the winter phase and to stay as strong as possible during this period by collecting sufficient stores to allow for successful overwintering.

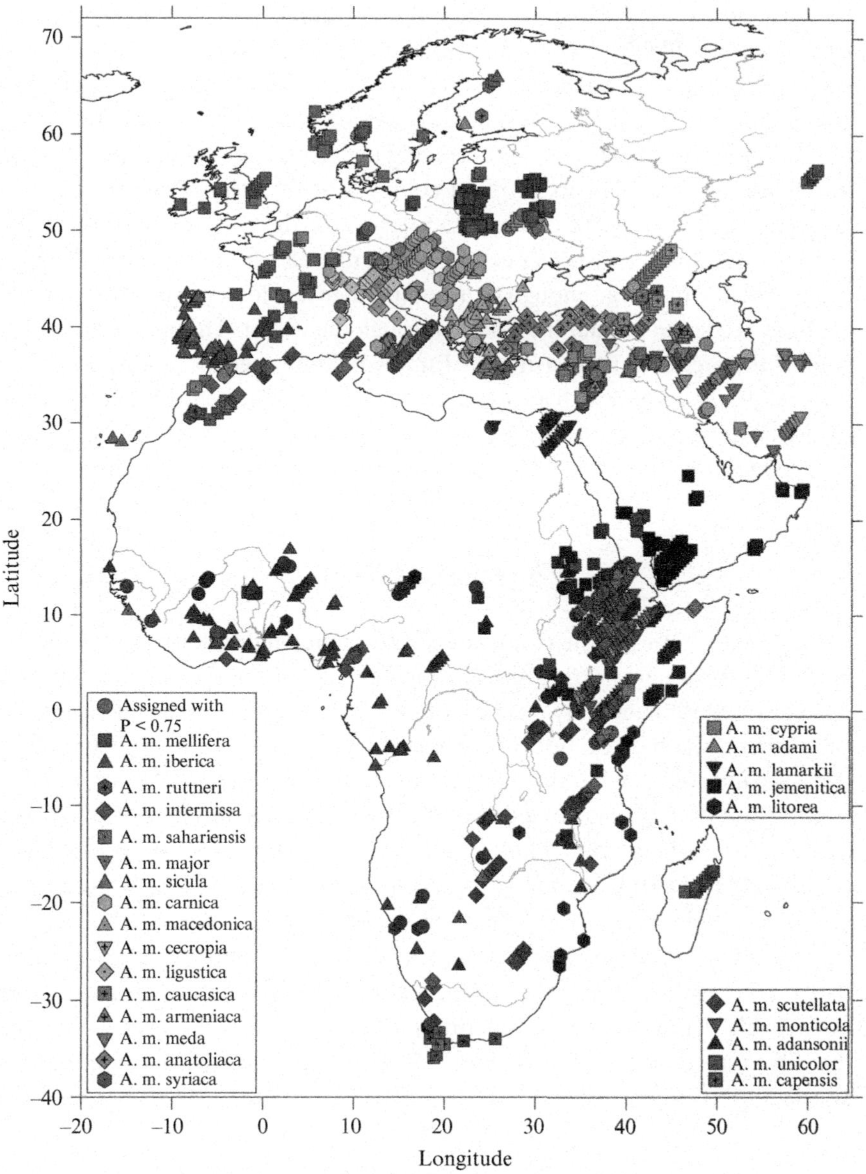

FIGURE 9.2 *The distribution of endemic A. mellifera subspecies identified based on their morphometry. Similar colors (grey shades) reflect similarity in morphological features (for a color version see tuck-in page between 142–143).*

But it is not size alone that matters. It is also the ratio between wintering bees and honey stores available. If winters are long, more honey will be needed per bee to keep the colony alive. Hence, colonies of more Nordic or Alpine subspecies are characterized by smaller wintering units compared to those in Mediterranean climates, where colonies can start foraging and brood production much earlier in the season. It is also about rapid colony growth after the winter phase to achieve a sufficiently large colony size to allow for the production of reproductive swarms. Yet all subspecies exposed to temperate climates must find appropriate nest cavities large enough to accommodate a colony of sufficient size and the food needed to sustain the colony through the winter. Thomas Seeley (1977) studied the preferred nest cavities of European honey bees in meticulous detail. Swarms prefer cavities with a volume of 40 liters, and they choose a new nesting site in the most democratic way by quorum sensing, as a series of brilliant studies revealed that are summarized in the book *Honeybee Democracy* (Seeley 2010).

9.4. Tropical honey bees

Honey bee colonies in the tropics can take a much more relaxed approach to searching for a new nesting site (Schneider and Blyther 1988; Schmidt and Thoenes 1990). They may also encounter periods of dearth, but whenever these occur, the bees do not respond with a hibernation phase. They simply leave the unfavorable habitat as a migratory swarm to search for a rewarding nectar flow elsewhere (Schneider 1990). Hence, the requirements for the size of the nest cavity are much less constrained. There is no need for insulation to protect the colony against subfreezing temperatures. There is no need to establish large stores for colony survival because new stores can be easily foraged at the new location. Colonies are much smaller, and many more swarms are issued in a year compared with the temperate climate honey bees. Table 9.1 shows the major differences in colony traits resulting from the different reproductive strategies, clearly demonstrating the typical r and K selection traits of tropical and temperate honey bee colonies, respectively.

TABLE 9.1 Typical Traits Reflecting the Different Ecological Strategies of Tropical and Temperate Honey Bee Colonies

Trait	Temperate	Tropical
Colony size	30,000–40,000	15,000–25,000
Number of swarms	1–3	5–10
Migratory/absconding swarms	None	Frequent
Cell diameter[a]	5.4 mm	4.6 mm
Drone production	Associated with swarming	Year-round

[a]Data from Martin and Kryger, (2002).

Indeed, together with Rodolfo Jaffe and Vincent Dietemann, we showed how this migratory behavior results in exceptional dynamic population composition in African honey bees (Jaffé et al. 2009). We monitored the genetic population composition in the nature reserve Telperion (Gauteng, South Africa), where we had studied the composition of a drone congregation area (DCA) over three years. More than 80% of the queens contributing drones to the DCA had been replaced by mostly unrelated ones within a period of only eight months. Hence, not just colonies but also entire populations of African honey bees are constantly on the move, which provides an insight into factors that allowed for the rapid spread of the African honey bees in South America after their introduction into Brazil.

9.5. Lessons from ecology

Ignoring regional adaptations that are the result of thousands of years of natural selection does not seem wise for any honey bee breeding program. We do not question the value of breeding per se; however, we do have concerns about breeding a bee that can serve as a jack of all trades. It will simply be impossible to breed a honey bee that performs equally well throughout the world. There is no such thing as a "perfect bee," and we advocate a very strong emphasis on regional bee breeding because what is good in one year may not be in the next. In classical husbandry, although some attention is paid to local conditions, the animals can be kept under (sometimes extremely artificial) standardized environmental conditions. The animals can only survive because of the support they receive from the farmer. In honey bees, this is fundamentally different. They are not moved in and out of a stable, a coop, or a den daily. Most important, however, given that they find a suitable nesting site, they are perfectly able to survive in the wild (if any suitable habitat is available). Moreover, they are not fed for growth. They are maintained in exchange for the honey the beekeeper removes from the colony. Thus, there is no such thing as a "domesticated honey bee." Bees are kept (that is why it is called beekeeping and not bee farming), occasionally with some moderate bee breeding, which has not profoundly changed the genetic identity of the original endemic subspecies in Europe, Africa, and Asia. Even in America (both North and South) and in Australia, where honey bees were introduced by European settlers hundreds of years ago, we can still identify the various subspecies that contributed to the colonization of the continents by feral honey bees (De la Rua et al. 2009). If bee breeding is going to have a sustainable impact both ecologically and economically, it is important to acknowledge that natural selection has shaped the huge diversity of honey bees in endemic distributions with colony dynamics that fit the regional climatic and foraging conditions. For example, beekeepers who transport a colony from southern

Italy to Norway should not be surprised if the colony dies in the second winter after producing huge honey yields in the first year. The Mediterranean climate is known to differ markedly from that of northern Europe. It has shaped a completely different flora and fauna over several thousands of years since the last glaciation. Clearly, the honey bees are an important part of the fauna and essential for the flora. It would seem wise to respect these results of long standing co-evolution, not just from a perspective of biodiversity conservation but also from that of honey bee health.

Several nonsustainable schemes have been proposed to eliminate less productive endemic races and replace them with introduced presumably "superior" ones. The most extreme example may be the replacement of *Apis mellifera mellifera* (the European black honeybee) with *Apis mellifera carnica* (Carniolan honey bees) in an attempt to improve the behavioral characteristics of managed honey bees in Germany. In a huge selection effort encompassing the vast majority of the German beekeeping community, bee breeders measured specific angles in the wing venation patterns of queens to eliminate *A. m. mellifera* queens and hybrids between the two subspecies from the population (Goetze 1940, 1959, 1964). The "cubital index" (Figure 9.3) in the wing venation pattern became the holy grail of honey bee breeding. Thus, if a colony performed well but the queen had too small a cubital index, it was eliminated from the breeding population (Ruttner 1988. Despite this complication, there was nevertheless measurable success in producing the desired behavioral traits (reduction in swarming, defensive behavior, and honey production), and the selected traits in the wing venation pattern changed in response to the extreme selection intensity. However, other morphological traits that had not been selected for remained remarkably *A. m. mellifera*-like, suggesting that substantial hybridization had been taking place despite the enormous effort to control mating (Moritz 1991b). Mitochondrial DNA studies have shown that managed colonies may have

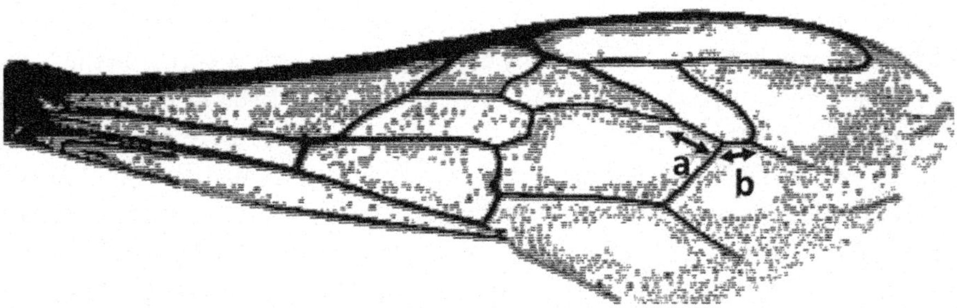

FIGURE 9.3 *The "cubital index" in the wing venation is computed by dividing cell branch a by branch b. A high index is an indication for a Carniolan-type honey bee, which had been preferred by many breeders aiming at A. m. carnica stock.*

had the desired breeding type; however, this tells us little about the composition of the nuclear genome (Pinto et al. 2014), which determines most of the phenotype. The mitochondrial genome is exclusively transferred through the queen lineage. Because this can be very accurately controlled in honey bee breeding, it is not surprising to see the C haplotype, of the Carniolan honey bee strains, as the most frequent in breeding programs with *A. m. carnica*. The male contributions may look much more heterogeneous depending on the stringency of mating control in the various programs. Breeding lines based on using honey bee-free North Sea islands might have had better control over the male side compared to those using land-based mating stations. However, even on the islands, matings with drones from the mainland cannot be completely excluded (Neumann et al. 1999). This leaves instrumental insemination of honey bees as the only means to confidently control mating, which has been used in very few honey bee breeding programs.

9.6. Lessons from natural selection

Many of the traits that are important to beekeepers are not favored by natural selection in honey bees. As discussed previously, beekeepers aim for very strong colonies with the more worker bees, the better. They are afraid of losing bees through colony fission and reproductive swarming. Hence, they not only select against swarming propensity but also implement management measures to prevent swarming. However, selecting for honey bees that do not swarm stops reproduction at the colony level, which is particularly maladaptive from an evolutionary perspective since it effectively neuters the colony. Obviously, the bee breeder will get little assistance from natural selection processes by pursuing such a trait as a goal of selection. The same may be true for defensive behavior. Predators of honey are highly abundant. We are not aware of any vertebrate that is not a keen consumer of honey if it is made available. Hence, it seems more than just prudent for a colony to have a highly efficient colony defense in place. Again, natural selection will select for defensive honey bees, a situation not desired by the beekeeper nor by his neighbors. Honey production, however, is in the interests of both the beekeeper and the colony. In this case of convergent interests, does this make it easier to select for this trait? Although this seems to be an advantage at first sight, the opposite is true. How to enhance honey production through breeding if this has already been selected for by natural selection? There will not be much selectable genetic variance in a population if the trait is under strong positive natural selection (Mousseau and Roff 1987). It will already have been optimized by natural selection within the genomic and ecological constraints. From a selective breeding standpoint, it is actually much easier to select for less defensive and nonswarming bees rather than for highly productive honey bees. However, we should also keep in mind that our attempts at selection may shape honey bees that are less well adapted to natural conditions.

Nevertheless, honey bee breeders have achieved some remarkable results, particularly with respect to swarming and defense behavior. Economically successful breeding lines in temperate zones are characterized by markedly reduced defensive behavior and an early and steep growth in colony size, resulting in larger colonies and hence larger honey yields per colony. The efficiency per worker is not necessarily enhanced; rather, the number of workers per colony that can be allocated to foraging increases honey yields. As a consequence, European honey bee subspecies south of the Alpine ridge have been favored in commercial bee breeding because they are adapted to an earlier start in colony growth after the winter phase compared to subspecies north of the Alps. Although this strong colony growth may be highly advantageous for migratory beekeepers that follow honey flows, it may be less optimal for colonies that remain at fixed apiary sites. Here, the nectar and pollen sources may not be in line with the demand by the bees, and colonies may need additional feeding to take full advantage of the additional growth potential of subtropical bees in a temperate climate. However, extended brood production has become a double-edged sword since *V. destructor* became a ubiquitous problem (Rosenkranz, Aumeier, and Ziegelmann 2010). Early brood production results in early mite production and hence requires more intensive mite control. All resistant honey bees in Europe show slow colony development, making them less attractive for beekeeping if honey production is the main goal. There seems to be a negative trade-off between the suitability of the honey bee for apiculture and *Varroa* resistance. This may be most obvious for African honey bees, which have been shown to be resistant not just to *Varroa* but also to most lethal pests and pathogens (Strauss et al. 2013). Their small colony size, unpredictable defensive behavior, and propensity to abscond and produce migratory swarms are more than just a problem for any beekeeping operation in industrialized countries with temperate or even subarctic climates. Dense human communities are incompatible with aggressive honey bees, and small colonies cannot survive the long periods of cold winter.

In general, it may not necessarily be a good idea to embark on a search for a bee that represents the best for beekeeping. Climatic conditions are extremely variable and the various types of colony management too diverse for a single best honey bee lineage to be identified. In fact, it has been suggested repeatedly that it is the extreme genetic diversity in the colony that is important for its fitness (Fuchs and Schade 1994; Mattila and Seeley 2007), although others have failed to find a strong effect of genetic diversity on colony efficiency (Neumann and Moritz 2000; Pirk et al. 2014). Whether or not genotypic variance is important for colony efficiency, it is clearly most relevant to facilitate division of labor among workers in the colony. We know that workers sired by different fathers specialize in certain tasks, such as defense or water collection, nectar, and pollen foraging (Page et al. 1995; Kryger, Kryger, and Moritz 2000). For the ideal colony, the queen needs to be mated with the right combination of males that carry genotypes that result in workers with optimally complementary skills (Fuchs and Moritz 1999). The right mix of workers is important, and the diversity among the drones with which the queen mates will be tightly linked to colony efficiency.

Moreover, what would we do if the best honey bee strain had been found? Would we really want to spread this throughout the world of apiculture? There is every reason not to search for the best bee on a global scale, because it has been repeatedly shown not to be an achievable goal. This has been best documented by trying to introduce "superior" European honey bees into African beekeeping. None of the recurrent and sometimes large-scale introductions have had any sustainable impact on the bee colonies used in Africa south of the Sahara (Hepburn and Radloff 1998). The imported honey bees were politically correct according to the nationality of the corresponding developmental aid program or the whims of the local beekeeping fraternity (Fletcher 1977). No one considered that natural selection might have shaped the regional honey bees to deal with local climatic and habitat conditions better than the imported ones. Imported colonies were quickly taken over by queens of regional subspecies. Wild migratory swarms can take over introduced colonies by simply killing the resident queen and usurping the introduced colonies before these even have a chance to produce any drones, let alone a reproductive swarm. The introduction of African honey bees into South America independently confirmed this result in an experiment on a continental scale, albeit with a small sample of 56 queens imported from Africa in 1956 into Brazil. In just over three decades, the African honey bees had spread throughout the continent, and they are now integrated as a sustainable member of the continental bee fauna in South- and Mesoamerica as well as in the southern US states (Schneider, Hoffman, and Smith 2004). Fortunately, this has had no detectable impact on the conservation of native bee species (non-honey bees), and only the introduced European honey bees seem to have been replaced by the spread of the African bees in America (Moritz, Härtel, and Neumann 2005). This example of the spread of an invasive alien species underscores the points we made previously about the selection of honey bee subspecies for local conditions. It also demonstrates the strength of natural selection in relation to the rather more modest efforts of honey bee breeders. Selection against nature requires exceptional efforts. Any breeding attempts in regions with strong wild or feral honey bee populations will probably be difficult at best if they rely on natural mating and are unable to control the locally abundant wild drones.

When apiculture is used to improve socioeconomic conditions in developing rural regions, these initiatives have the primary goal of implementing sustainable colony management rather than sophisticated breeding attempts. Globally, the majority of honey bees are kept in small apiaries run by hobbyists or as a source of additional income in small-scale farming. In these small apiaries, it is not important whether a colony produces 20 kg of honey instead of the 22.5 kg that may be achieved from a highly bred top colony, but it is important that the colonies are adapted to local conditions.

9.7. *Apis mellifera* in Asia

Whereas European honey bees were imported into the American and Australian continents with no endemic *Apis* species, *A. mellifera* has also been massively imported to eastern Asia for honey production, where all the other species of *Apis* are endemic. The example of China and Vietnam shows how honey production has soared impressively in the past decades through the use of *A. mellifera* in apiculture (Pirk, Crewe, and Moritz 2017; Figure 9.4). *Apis mellifera* was first introduced into China in 1920 but has seen a huge growth since then. Today, the apicultural population is dominated by the western honey bee in most of China, and *Apis cerana* only survives in the wild, where it is not affected by beekeeping with *A. mellifera.*

This provides a very different scenario from that of other continents on which *Apis* is not indigenous, with many problems peculiar to the interactions of *A. mellifera* with the other native *Apis* species. The biology of Asian honey bee species has been comprehensively dealt with in the

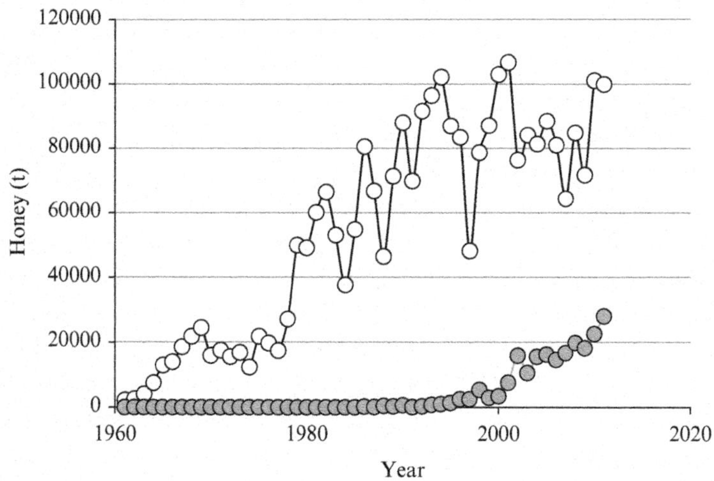

FIGURE 9.4 *The increase in honey production (in tons) in the People's Republic of China (open circles) and Vietnam (shaded circle) as reported by the Food and Agriculture Organization of the United Nations. Note the exponential growth in Vietnam starting in the late 1990s.*

treatise edited by Randall Hepburn and Sarah Radloff (2011). The Asian *Apis* species comprise the open-nesting dwarf honey bees *Apis florea* and *Apis andreniformis*; the cavity-nesting *A. cerana, Apis negrocincta, Apis nuluensis,* and *Apis koshevnikovi*; as well as the open-nesting giant honey bees *Apis dorsata* and *Apis laboriosa*. The cavity-nesting species are very similar to *A. mellifera*, and indeed *A. cerana* is phylogentically the closest sister species to *A. mellifera*. Both species have many similarities not just in morphology but also in many aspects of behavior. Even interspecific matings between the two species can occur, although no offspring will be produced even though the sperm can fertilize the egg. It is only after the blastoderm stage that embryonic development is aborted (Ruttner and Maul 1983). Colonies can even operate as mixed species units by mixing brood combs of both species in the same hive or introducing *A. mellifera* queens into queenless *A. cerana* colonies. This high level of similarity sets the stage for potentially strong interspecific competition, with various reports of *A. cerana* queens having reduced mating success if *A. mellifera* drones are in the vicinity (Ji et al. 2003). Nevertheless, natural selection seems to help, and wild *A. cerana* populations appear to be well established in many regions of Asia (Abrol 2013). One reason for this resilience may be the *Varroa* mite, which is lethal to *A. mellifera* but not to *A. cerana*. In *A. cerana*, the mite can reproduce only in the drone brood, whereas *V. destructor* reproduces in both worker and drone brood in *A. mellifera,* causing colony collapse. However, the story of the interaction between *Varroa* and its two host species may not have been fully revealed. Recently, new types of *Varroa* have been identified, showing multiple host shifts (Roberts et al. 2015) but also highly specific types for *A. cerana* and *A. mellifera*, even if the colonies are kept in the same apiaries (Beaurepaire et al. 2015). Therefore, factors other than *Varroa* mite infestation may contribute to the selective disadvantage of *A. mellifera*.

Reports of potential direct competition between *A. mellifera* and the endemic Asian honey bees are as diverse as they are inconclusive. Whereas some claim dramatic losses of *A. cerana* after the introduction of *A. mellifera* (Sakai and Okada1973), others show the opposite (Koetz 2013a). *Apis cerana* and *A. mellifera* did not coexist successfully on the Solomon Islands, where managed *A. mellifera* colony numbers declined severely and honey production ceased entirely after *A. cerana* was introduced to the islands in 2003 (Anderson 2010; Anderson et al. 2012). It is likely that all reports are true depending on where and when the corresponding studies or observations were made. *Apis cerana,* like *A. mellifera,* has adapted to a multitude of climatic conditions, from temperate to tropical. The only general pattern that emerges from the introduction of *A. mellifera* into Asia is that transhumance has resulted in a major global problem for beekeeping that has now been in existence for more than four decades: the spillover of *Varroa* mites from *A. cerana* to *A. mellifera*. It seems as if only a few specific lineages of *Varroa* had the potential to swap host species from *A. cerana* to *A. mellifera*. Today, two lineages of *Varroa*, the Korean haplotype and the Japanese haplotype, can be found in *A. mellifera*, requiring all beekeepers working with European *A. mellifera* to regularly treat their colonies (Rosenkranz et al. 2010). The bleak reality is that the beekeeping industry in Asia has not learned its lesson from this disaster. Whereas colony numbers are in decline in North America and western Europe, Asia has seen a tremendous growth in the number of managed *A. mellifera* colonies. In China alone, more than 9 million colonies were reported to the Food and Agriculture Organization in 2013—three times the number of colonies reported in 1961—showing the exceptional growth of apiculture in this region (Moritz and Erler 2016). The continuing increase in *A. mellifera* apiculture in Asia not only perpetuates the *Varroa* spillover problem for the rest of the world but also puts the Asian endemic honey bee fauna at risk. It has been suggested that pests and pathogens also spilled over from imported *A. mellifera* to native *A. cerana* colonies (Abrol 2013). For example, European foulbrood and acarine disease (*Acarapis woodi*) have transferred to Asian honey bee species probably through the importation of *A. mellifera* colonies (Allen, Ball, and Underwood 1990; Abrol 2013; Forsgren et al. 2015).

It may only be a matter of time before the first African honey bee colonies arrive in Asia, with the likelihood of a much more dramatic impact on the native honey bee species than European honey bees have had. Fortunately, it seems that the *Apis* species primarily compete among themselves and not with all the other social and solitary bees. The spread of African honey bees in America did not lead to known species extinctions of the rich bee fauna of the American tropics. They are very likely to spread as they did in the Americas, with the only difference that they might well outcompete the endemic Asian honey bee species with a real risk of the reduction of the genus *Apis* to a single species.

9.8. Disease transmission by apiaries

Apiculture is based on apiaries in which colonies are placed next to each other in close proximity for the convenience of the beekeeper. Under natural conditions, these excessively high densities of colonies rarely occur in the cavity-nesting species. An exception may be the nest aggregations of the open-nesting giant honey bees in Asia. They form dense and highly conspicuous colony clusters on so-called bee trees where many colonies are aggregated on the same tree (Figure 9.5).

However, the cavity-nesting species need to search for cavities that are dispersed and highly cryptic. These species tend to prefer concealed nesting

FIGURE 9.5 *Nest aggregation of Apis dorsata on a bee tree.*

locations to avoid predation. For *A. mellifera,* the apiary may be the major problem associated with being kept by beekeepers. Putting bee colonies in a box, providing them with combs in movable frames, housed in various supers, is less of an interference of honey bee biology than the densely packed colonies in the apiary setting. The rows of colonies at a high density pose a serious orientation problem for the members of the colony; after returning from a flight, all queens, drones, and workers must identify their home colony among the multitude of hives placed in the apiary (Figure 9.6). Therefore, beekeepers often aid bee orientation by using different-colored hives. This is known to help both the beekeeper and the bees quickly identify the colony. Nevertheless, all apiaries are known to foster "drifting" of workers (Rauschmeyer 1928).

Honey bees get lost more frequently during their first orientation flights and may end up in neighboring colonies. The first and last colonies in a row of hives tend to accumulate drifting workers (Jay 1966, 1968, 1969; Neumann, Moritz, and Mautz 2000). Although this may seem quite irrelevant from a honey production standpoint, because the bees are not lost from the apiary, it is highly relevant for the transmission of diseases among colonies. Horizontal transmission of pathogens among colonies is greatly facilitated by crowded apiaries (Seeley and Smith 2015), and at the same time this configuration of colonies will also select for more virulent pathogens (Fries and Camazine 2001). Pathogens confront a trade-off between transmission and virulence, with higher efficiency in transmission often resulting in reduced pathogenicity. Beekeeping provides for transmission at no cost to the pathogen, thus allowing for the selection of more virulent types. Moreover, drifting of queens can be lethal for a new host colony if a drifting queen gains access and kills the resident queen. It can also be fatal for the drifting queen if she is killed in the new host colony. Whatever the outcome of these interactions, the queen's home colony is placed in jeopardy because it is now hopelessly queenless.

Drifting of drones is considered to be an efficient parasitic strategy (Rinderer et al. 1985) because they seem to be more readily accepted by other colonies than are queens or workers. Maintenance of drones in host colonies imposes a resource burden that is not compensated for through the transmission of host genes.

FIGURE 9.6 *Master beekeeper Georg Klindworth in his skep apiary. This is a classical skep apiary in which honey bees easily return to the wrong hive because all colonies are placed in close proximity. This is a setting that is likely to enhance the transmission of diseases, the chances of robbing, and the loss of queens returning to the wrong colony after mating flights.*

There is increasing evidence that the pathogen load in large-scale apicultural operations is markedly higher than in colonies from small apiaries or even among wild colonies. For example, twice as many colonies are lost by large-scale migratory beekeepers compared to stationary ones in South Africa (Pirk et al. 2014). In a study screening for viral infections, the pathogen load in large-scale beekeeping operations was twice as high as that of small-scale stationary apiaries, which in turn had a pathogen load twice as high as that in wild colonies. Hence, the viral pathogen load is increased by almost an order of magnitude. These empirical findings very decisively confirm concerns about the effects of apicultural management practices that only reflect standard epidemiological theory and inevitable processes in which pathogen transmission is facilitated.

9.9. Hives with movable frames

Honey bees are known for their efficient nest construction techniques. Beekeepers assist their bees by providing hexagonal cells on comb foundation to the hive that ensures the most efficient use of the available space. As discussed in Chapter 6, the bees themselves are less reliable builders. For a more detailed review of honey bee nests and comb usage, see Hepburn, Pirk, and Duangphakdee (2014). Hexagonal cells, however, are not specific to honey bees and stingless bees; various eusocial wasp species also use the same format for cell construction. Whereas cells in wasps' nests constructed from wood pulp are truly hexagonal, those of honey bees are often not. In fact, honey bees are actually rather sloppy architects and often deviate from regular optimal patterns of cell construction. What is specific to honey bees is the vertical and parallel wax comb construction. Even if the beekeeper does not provide a frame for comb construction, honey bees make almost parallel combs when building a nest in cavities, and they also do so in the wild (Figure 9.7). Another specialty of honey bee combs is that the same type of cell is used for storage of honey and pollen and also for rearing the worker brood. Other social bees use the hexagonal cells only for rearing brood but not for storing food. The combs are not only important for what they contain (brood and stores) but also provide the substrate on which the adult bees

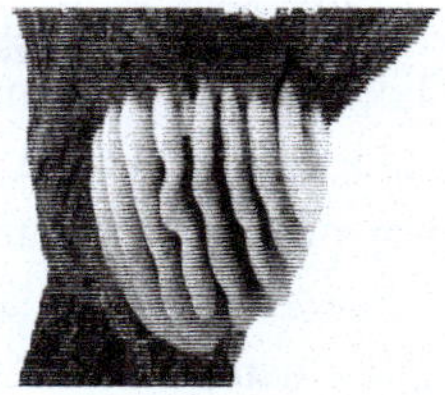

FIGURE 9.7 *Combs of a wild colony built almost in parallel. Movements of workers over the tops of the combs are not possible because they are firmly attached to the substrate.*

reside. The vertical structure of the combs imposes a constraint on the complex communication system of the dance language, where the direction to a food source cannot be directly indicated but needs to be transposed onto the vertical surface of the comb. The sun's position is indicated by the direction of the dance on the vertical surface, with the angle between the sun and food source indicated by the deflection of the dance from the vertical position. In open-nesting dwarf honey bees, the workers indicate the direction of the food source directly on the horizontal dance floor on the top of the single comb of the nest, and no transposition is necessary as in other honey bee species. The vertical structure of the comb imposes additional levels of sophistication on the communication system that has evolved to provide for efficient recruitment of foragers. However, this is not the only problem that requires resolution. The vertical combs are rather effective barriers to the movement of bees from one comb to another. Whereas in a beekeeper's box they can at least choose to move over the top or the bottom of a frame, in the wild they can only use the bottom or the sides but not the top.

Workers have to walk quite a distance to move from one side of the comb to the other. Whereas this costs time and energy in the summer season, it becomes a functional problem in winter when outside temperatures drop below freezing. The workers can actively heat the brood nest by shivering their flight muscles, and the energy required to generate this heat is almost as high as that required for flight (Southwick 1985). The bees use their honey stores as fuel to produce the heat required to maintain colony temperature.

Although the colony can maintain a temperature well above 30°C in the core of the brood nest, this is not true throughout the winter cluster. Bees at the periphery can cool down, fall into a chill coma, and ultimately freeze if outside temperatures fall far below 0°C over extended periods of time. The compartmentalization of the brood nest by the combs may now turn out to be a lethal trap. The bees do not form an undifferentiated cluster as they do in a swarm, but the cluster consists of slices of bees caught between the combs in the nest. As it cools down, it will become increasingly difficult to move from one side of the comb to the other. Most models of thermoregulation assume that the bees move freely from the warm core of the cluster to the cooler periphery. Yet this is only easily possible within the confined space between two combs. Although the colony does form a spherical cluster of individuals that is ideal for temperature regulation, it is composed of slices that are very small on the periphery and hence more prone to freezing than the larger slices in the center of the cluster. The stores of fuel are in the periphery of the colony, and this makes it difficult for the bees to achieve a compromise between the availability of honey and a suitable position for the thermally stable cluster of workers. Once the honey stores are consumed on a comb, it is no longer a suitable place for the winter cluster to remain. They need to move to the next comb, which, if the bees are fortunate, will still have unused stores. The movement from one comb to another can become a major risk for the colony. It is not uncommon for a beekeeper to find starved colonies after winter that simply lost contact with their honey stores. The solid structure of the comb becomes problematic for thermoregulation in winter. Clearly, it would be highly adaptive if the bees constructed small openings in the center of the combs to allow for shortcuts to get from one side to the other. Although they clearly have the ability to construct such passageways, we are not aware of any reports in which such modifications of the comb have been observed in a honey bee colony. It appears that the cell construction behavior of the building workers precludes building horizontal passageways through the nest. If comb construction were regulated in a more self-organized manner, the bees might have been able to use a cleverer design that overcame the deficiencies exposed during winter, with its need for energy-intensive thermoregulation.

In fact, honey bees might be better off in a beekeeper's colony with a brood chamber and honey super rather than in a natural nest. The major revolution in keeping honey bees was made possible by a hive design that was described in a patent filed by Lorenzo Langstroth on October 5, 1852 (Figure 9.8), which used

FIGURE 9.8 *The Reverend Lorenzo L. Langstroth (1810–1895) patented the movable frame hive for apiculture.*

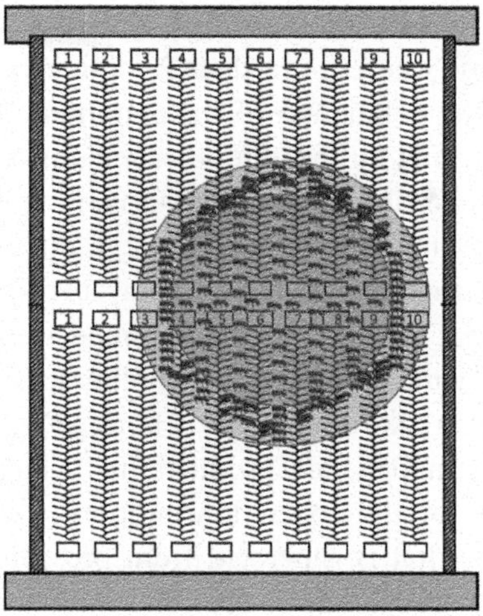

FIGURE 9.9 *Distribution of workers and honey in a winter cluster in a hive. The space between the two supers allows for exchange of workers in the center of the cluster. In a natural colony, this is not possible because bees build only one set of parallel combs and there are no central passageways allowing for the movement of bees from one side of the comb to the other.*

movable frames that could be managed from the top, had several supers, and crucially described the "bee space" (Kritsky 2010). This allowed beekeepers to use colonies with interchangeable parts that consisted of several stories, each with its own set of frames.

This is a very different configuration of the nest than that established by colonies in the wild. The distance between the frames in the supers allows the bees to move more easily as they follow the honey stores they consume during the course of the winter. We do not know what price overwintering colonies pay under natural conditions. However, colony losses in temperate climates occur most often during the winter period, even when they are kept in Langstroth-style hives, in which the beekeeper provides the colony with some shortcut routes between the frames of the brood chamber and the super (Figure 9.9). We suspect that winter mortality is much higher in colonies nesting in natural cavities than in those in hives in apiaries.

9.10. Migratory beekeeping

Often, beekeepers do not just have their hives located at particular apiary sites but also take their colonies to new sites to follow rewarding honey flows. Although this may be economically important for a beekeeper's business, migrating colonies enhances the problem of spreading pathogens and parasites from the level of a single apiary to the regional or sometimes even continental scale. Particular massive migrations occur when beekeepers are asked to provide honey bee colonies for pollination services. The most extreme case is that of the Sacramento Valley, where more than 2 million colonies are transported every year to ensure the pollination of the almond orchards. Large-scale pollination is also a rewarding business in other regions of the world where large fruit orchards require reliable pollination services. Indeed, natural and wild pollinators are at an extremely low abundance in intensive plantation agriculture. Often radical destruction of natural habitats in combination with chemical pesticide treatments that are characteristic of intensive agricultural production have created a strong demand for managed pollinators. Large commercial apicultural businesses often take advantage of both sources of income: provision of pollination services and honey production. In fact, many migrate to regions of rewarding nectar and pollen flow in order to establish strong colonies for their main income that is generated from providing pollination services. Hence, extremely large truckloads of bee colonies are transported over several thousand kilometers. In regions where honey bees are not endemic, this is an issue with regard to colony health because pathogens and pests are instantaneously dispersed on a national and sometimes continental scale. In regions with endemic honey bees, this raises a further problem with serious biodiversity implications whenever subspecies are transported from their endemic region into a region where other subspecies are endemic (Soland-Reckeweg et al. 2009). This causes the genetic admixture of subspecies that have been evolving over thousands of years. Because natural selection has shaped these subspecies, this admixture is unlikely to improve regional stock, and it certainly puts the success of regional breeding schemes at risk and requires increased efforts to control mating.

The worst example of the detrimental effects of long-distance migration derives from Africa, where migration has resulted in the second largest instance of colony losses in the world, resulting from ignorance and inappropriate colony management. The so-called "capensis calamity" is a story that exemplifies one of the darkest moments in apiculture (Allsopp 1993; Allsopp and Crewe 1993). Migratory beekeepers providing pollination services in the Western Cape province in South Africa moved some of their colonies approximately 2000 km northeastward to take advantage of the Aloe honey flow in the Highveld region near Rust de Winter. Their goal was to double the number of colonies by splitting them into two. Most commercial beekeepers in the northern region of South Africa followed this routine, with as many as 13,700 colonies being set up to benefit from rapid colony growth early in the season (Steinhobel 1976).

South Africa harbors two distinct subspecies. The first, *Apis mellifera scutellata*, is a widely abundant honey bee race with a distribution range from the Karoo in the south throughout eastern Africa south of the Sahara. The second subspecies, *Apis mellifera capensis*, is endemic to the fynbos biome at the southern tip of the African continent (Onions 1912; Figure 6.13). This subspecies is very different from all other honey bees, as we have detailed previously. One trajectory of their life history is the production of parasitic workers (Neumann and Moritz 2002). The specialty of the parasitic workers is to enter queenless foreign colonies and establish themselves as false-queens. Splitting of colonies into one queenright and one queenless unit provides ideal conditions to activate the parasitic worker trajectory of *A. m. capensis*. Because hundreds of *A. m. capensis* colonies had been brought to Rust de Winter, workers from these colonies readily took over the queenless units, produced more parasitic worker offspring that also invaded queenright colonies, and caused a loss of more than 80,000 colonies within a few months. It took South African beekeeping many years to recover from the impact of this consequence of migrating colonies over long distances, and a lineage of parasitic workers persists to the present time throughout the region where managed *A. m. scutellata* colonies are used by beekeepers. Only strict regulations on bee transport could re-establish beekeeping on a commercial scale. A line dividing the country from the Atlantic Ocean to the Indian Ocean was defined, across which transport of honey bee colonies was prohibited (Minister of Agriculture, Forestry and Fisheries 2013). This has reduced but not eliminated the problem.

9.11. Colony collapse disorder

Colony collapse disorder (CCD) achieved celebrity status at a global level in the public media while relatively little attention was paid to the capensis calamity. CCD accounted for the largest loss of managed colonies in 2006 in the United States (van Engelsdorp et al. 2007). More than 2 million colonies were lost within one season after having been used to pollinate the almonds in California. Many media reported this as being close to 90% of the total number of commercially used colonies in the United States. The symptoms were completely new to honey bee pathologists. Workers would simply disappear, leaving beekeepers with empty boxes. Attempts to identify a specific pathogen or pesticide responsible for these symptoms failed to provide any conclusive evidence (Cox Foster et al. 2007). Many scientists therefore agreed that CCD is the result of a combination of different factors, including pests, pathogens, and pesticides (van Engelsdorp et al. 2009). If correct, it is a somewhat unsatisfactory conclusion because it does not help the beekeeper do anything specific. Whatever caused CCD, it did not recur on a similar scale in subsequent years. Colonies still die with CCD symptoms, but the numbers are much lower. So, is the problem no longer real? We think not. Not knowing what causes the symptoms means it can happen again at any time. What is certain, however, is that small-scale beekeepers were much less affected than their large-scale commercial colleagues. We are not aware of any reports that feral colonies, particularly the Africanized ones in the southern US states, were affected in any way. There is every reason to suspect that the industrialization of beekeeping operations per se may have been a major cause of the observed colony losses. If 2 million colonies are concentrated in a relatively small area, virtually every know honey bee pathogen will also be present in that area. Drifting of workers among colonies will be extreme, and so will the spread of pathogens. Practically every combination of virus, bacteria, and fungus will be possible in such a setting, and it is an ideal stage for the production of multifactorial colony collapse. We do not know if this is what happened, but it may very well be that putting 2 million colonies into the same valley is simply not a sustainable way to manage bees.

9.12. Conservation of honey bees

If invasions and introductions threaten pristine endemic honey bee populations, what can we do to conserve them? Unfortunately, this dilemma is one of the bleakest of all the dark perspectives of beekeeping. Regulations can be drafted and implemented; however, there needs to be a realistic chance that those involved in beekeeping will adhere to these regulations. This is the difficult part because beekeeping is not just a fun hobby but also needs to pay for people's livelihoods. Not much can be done that is going to be effective because human societies are obviously not set up to care about honey bee diversity. There are good reasons for this, because as humans, our foremost concern is for our own conservation and well-being—at least in the short term. The question is at what level do we need to conserve the world to achieve our own conservation in a sustainable way, and this is a huge societal conundrum that we certainly will not and cannot resolve in a volume on the dark side of honey bee biology. Yet any conservation of honey bees will interfere with human activities because honey bees are not only transported by beekeepers or eager scientists for particular purposes but also accidentally translocated as swarms on boats, trains, trucks, and even airplanes, as in the case of the introduction of the dwarf honey bee to Sudan. Because our societies are unlikely to return to simpler modes of transport, it is very clear and inevitable that despite all the regulations, eventually a swarm of *A. florea* will reach the Australian continent following in the footsteps of *A. cerana* that moved from Asia to Queensland in 2007. It is also clear that an African *A. mellifera* colony, now successfully selected to be resistant against *Varroa*, will eventually make it to some port in Asia and escape to the mainland in an unfortunate event. Import regulations do not prevent this from occurring; however, they will extend the time before it happens, and they will certainly reduce the frequency of intentional introductions of bees from one continent to another, which has been possible in the past. This indeed is an important contribution to postponing possible invasions. At the same time, the world is expected to change in climate, in land use, and in the number of people living on it. Clearly, this is also going to have an impact on the number of managed and wild honey bee colonies. The colonies will move in a northerly direction into what are still arctic regions, as they did from glacial refuges during the postglacial periods. Honey bees are perfectly adapted to dry savannahs and semideserts, so they will be less affected if the global warming scenarios result in the expansion of the Sahel region. In fact, they may become increasingly important for human survival in these regions because they produce a highly valued food product without large investment and a low requirement for water.

Thus, honey bees in general may be among the least affected organisms in terms of conservation and global change. Nevertheless, an important condition is that we manage to maintain the diversity of endemic subspecies. They provide the huge reservoir of specifically adapted honey bees for most climate zones other than the arctic. Honey bees do well in deserts, savannahs, cold winters, and periods of extended drought. They manage in the tropical rainforest but also master timber-producing conifer monocultures. As vegetarian generalists, they have been shaped by natural selection to fit a multitude of ecosystems. Once established in the wild, it seems little can be done to stop them from spreading. The ecological and climatic conditions will set limits to their distribution, which has not been achieved by eradications programs, poison baits, bee-free barriers, drone trapping, or breeding programs. These are the lessons that we should have learned the hard way from the experience of the expansion of the African honey bee in the Americas, *Apis florea* in Africa and the Middle East, and most recently *A. cerana* in Australia.

Nevertheless, or probably because of the invasion potential of some subspecies of honey bees, we should use whatever means we have at our disposal to support the conservation of native honey bee faunas. They have been shaped by natural selection to be adapted to the regions in which they occur. They are likely to be the species with the best chance of outcompeting potential invasive honey bee species/subspecies. Local adaptation, not the search for the best bee, will be the key to sustainable beekeeping. Preservation of local adaptation can be done by arranging bee breeding cooperatives, running regional selection programs, and promoting honey produced by regional bees. There have been several successful initiatives to conserve and/or restore original endemic bee races in Europe. Foremost are the Italian breeding programs to promote *Apis mellifera ligustica*. Similarly, there is a huge community of breeders working with *A. m. carnica*; unfortunately, not all of them are in the native region of this subspecies. Several breeding initiatives focus on the so-called black bee, *A. m. mellifera*, in northern and western Europe. There have also been attempts in Northern Africa and the Middle East to conserve endemic honey bee subspecies. In all cases, the goal is to establish regional reserves to maintain breeding populations. This can be done by providing beekeepers in these regions with queens of the appropriate ecotype. This is not very costly but is most suited to actively engage and involve the regional beekeeping community. There is no need for sophisticated mating control, and queens mated within these regional populations will inevitably be mated with drones produced by the queens provided by the program. Beekeeping will need to become more regional and profoundly less global in order to allow for a sustainable strategy to conserve honey bee biodiversity. And no, we are definitely not anti-globalization activists. Many enterprises clearly will not be sustainable at all if they do not have a global outreach. But the honey bee colonies of the world will remain under threat if the commercialization of beekeeping follows the path of the past decades, with a global bee market including trans- and intercontinental transfer of bee colonies—a quite unnecessary dark side of beekeeping. It is not just pesticides and pathogens, parasites and pests, but also human commerce and the increasing industrialization of apiculture that produce large-scale colony losses.

In essence, honey bees have not been successfully domesticated, but they have been very successful in establishing themselves in the wild. We have been able to identify a multitude of distinct subspecies whose diversity may reflect the local adaptations of honey bees to their endemic regions. The world's largest reference collection of honey bees, maintained by the Bee Research Institute of the University of Frankfurt in Oberursel, is primarily based on samples of workers taken from beekeepers' hives. Nevertheless, the subspecies can be easily identified, suggesting that these phenotypes reflect the biogeographic distribution of natural populations even though they are managed to different degrees (see Figure 9.2). Understanding the role of these adaptations for their regional success and understanding the implications both for the conservation of genetic diversity and for the practice of apiculture are the challenges faced by the beekeepers of the world. In order to ensure that future generations can enjoy and appreciate the lessons of the hive described in a host of books on honey bees, it will be necessary to manage honey bees with an understanding of their biology as the paramount guide to future practice rather than anthropomorphic conceptions of what an ideal bee should be. If, in the end, we are to conserve the bright side of the hive, it might be prudent not to move honey bees across continents but, rather, try to work with the bees we have right on our doorsteps for as long as we have them.

{ 10 }

Dark Sides of Honey Bee Science

Having shown how apicultural practice may interfere with honey bee biology in Chapter 9, the question arises as to how honey bee scientists approach these dilemmas. Does honey bee science benefit apicultural practice or the conservation of honey bee species? There are numerous stories of "mad professors" doing weird things that may eventually destroy the world, though fortunately most of them turn out to occupy the realm of fiction. Nevertheless, in the real world, scientists are often suspected of being inhabitants of ivory towers where their attention to relevant questions is distracted by their theoretical musings. We will not comment on the quality of honey bee research in general, since that would be invidious, and would require us to reflect on the quality and relevance of our own research that may reveal us to be inhabitants of the ivory tower without being aware of it. However, most researchers are driven by an urge to resolve open questions, although not all questions are equally important. Indeed, looking back on our own research, there have been a few occasions when we had the belief that our research mattered to the beekeeper. But this very personal impression may appear very different from the viewpoint of the beekeepers (for better or for worse). So perspective and personal interest are essential to determine the value of a specific research result, and we are happy to leave the assessment of the quality of research to others. In addition, scientists inevitably make errors, and these errors will sometimes be published in prominent journals, sometimes conflict with the truth, and sometimes conflict with another error. In the latter case, this will result in long-lasting controversies. Yet, this is actually the best of all scientific worlds because then issues are dealt with in a comprehensive way with a good chance that a better understanding might eventually materialize. Of course, honey bee science is prone to errors like any other discipline, and the course of our accumulating knowledge has been anything but straight. The most prominent example from honey bee research may be the controversy regarding the dance language. Karl von Frisch had meticulously deciphered the

FIGURE 10.1 *(Left) Karl von Frisch (1886–1982) received the Nobel Prize for discovering the dance language of honey bees. (Right) Adrian Wenner challenged von Frisch's theory by showing that odors are important for forager recruitment.*

dance language (von Frisch 1967), yet Adrian Wenner (reviewed in Wenner and Wells 1990) was not convinced and suspected odor cues were much more important for communication (Figure 10.1).

The issue has now been largely settled, with most researchers in the field acknowledging that bees do communicate rewarding foraging sites by a dance language, but this does not exclude the use of odors to assist in finding the sites (but see Wenner 2002). The controversy in and of itself (the presentation and defense of conflicting theories) is not at all a dark side of research; rather, it is the established procedure for eliminating errors and reconciling seemingly conflicting theories until new experiments provide different insights. However, the fierce battles among scientists behind the scenes are a different issue altogether and can sometimes escalate into undesirable conflicts. We as authors are actually quite happy to have been able to largely (but not completely) escape such dramatic events.

Here, we deal with the impact of science not in scientific journals or citation statistics but, rather, on the honey bees. Indeed, although we know a lot about the biology of honey bees, any impact of scientific knowledge on the actual life of honey bees on this planet has been rather modest.

Although we know the full genome (Honeybee Genome Consortium 2006) of honey bees and can make transgenic honey bees (Schulte et al. 2014), and are able to breed them with complete control of their genetic constitution, and we know how they forage, learn, and navigate and we understand how the colony operates and that it has wisdom, a spirit, and a democracy: these facets of knowledge are about understanding honey bee biology but not at all about changing the honey bee colony. When kept in a hive, the honey bee colony is arguably the best understood social insect in the world, but as outlined previously, humans have failed to develop truly domesticated honey bees. So far, humans have used science to understand but not to modify the colony (and we are actually quite happy about this).

Nevertheless, we present an example in which science mattered a great deal, not just for apiculture but also for honey bee biology. It is a story of an exceptionally logical, apparently outstanding scientific idea—one of those sound, knowledge-based decision that discerning policymakers love to have presented to them. Science translated into the real world for the benefit of the stakeholder, in this case the apicultural industry. It then turned out to seemingly be a major ecological catastrophe unnerving all the policymakers and not just terrifying the stakeholders but also the public and the media. However, it eventually became a formidable economic success for beekeeping, with the very media that had panicked initially paying scant attention to this success or acknowledging the impact of their initial trepidation.

This most famous example of the impact of honey bee science on apiculture was arguably that initiated by Warwick E. Kerr (Figure 10.2) when he started a project to determine whether tropical honey bees would do better in South America than the European honey bees that had been used to establish apiculture there.

Apis mellifera is not native to the Americas and had been imported by the early European settlers. Although these honey bees were producing

FIGURE 10.2 *Warwick E. Kerr (in his early eighties) in December 2005 when receiving his degree of doctor honoris causa from the Universidade Estadual de Campinas.*

reasonable amounts of honey in North America, it seemed that the colonies were not optimally adapted to the tropical and subtropical conditions of South and Central America (Michener 1975; Taylor 1977). After a trip to Africa, Kerr noted that the wild honey bees of Africa appeared to be well adapted to tropical conditions, and this shaped the idea of using them for honey production in the tropics of South America. To test their performance in tropical America, he successfully imported 56 queens from Africa to Brazil in 1956 with the aim of using a tropically adapted honey bee instead of maladapted European ones. The disaster started when colonies escaped from the quarantine station into the wild and spread into managed apiaries (Nogueira-Neto 1964; Kerr 1967). Beekeepers in the first decades after the introduction had no experience with highly defensive honey bee colonies and were taken by surprise by the mass attacks of African honey bees. As a consequence, there were many lethal accidents involving honey bees, and they were quickly called "killer bees" by the public media to differentiate the novel introduced African honey bees from the more benign European bees. Killer bees appeared frequently in the movie theaters and on TV, including the films *Killer Bees* (1974), *The Swarm* (1978), *Killer Bees!* (2002), and *Die Bienen—Tödliche Bedrohung* (2008), which enhanced the public's awareness of bees and their behavior.

The fictional stories about "killer bees" raised public awareness to a high level, resulting in concerns about global colony losses and refocusing public perception of the risk of losing honey bees altogether. In the early 1970s, colony numbers in Brazil were in decline in response to the introduction of African honey bees, falling to almost half the number recorded in 1961—a catastrophe for beekeeping. A similar pattern was also seen for honey production. This decline was much more dramatic than the drop in colony numbers observed in the United States that led to the global media hype about colony losses in 2006 due to colony collapse disorder. Imagine if you were the scientist being blamed for such an impact on beekeeping a decade after you had introduced the bees. Just picture yourself as Warwick Kerr: What would you have done in such a situation? How to repair the fatal error? As a dedicated scientist, Kerr searched for solutions, and he noted that production per colony had actually increased in the African type (Kerr 1967). Thus, the decline in honey production was not due to poorly performing bees but, rather, an effect of the decrease in apicultural activity. The industry had to learn how to handle the more aggressive colonies and adapt management procedures to deal with the intensive stinging behavior of the newly introduced bees (DeJong 1996). Apiaries were moved farther away from human settlements. The public became more alert regarding the risk associated with honey bee colonies. Nevertheless, it took almost a full decade after the introduction of the African honey bees to reach the original number of colonies. The developments of the subsequent decades vindicated Kerr, and honey became an export commodity in Brazil at the beginning of the millennium. Kerr was correct in surmising that the tropical honey bee would indeed outperformed its European counterpart by far. The Food and Agriculture Organization's database on honey bee colonies and honey production confirms Kerr's success with solid statistics. Since the introduction of African honey bees, the number of managed colonies has tripled in Brazil; honey production increased from 7,749 tons in 1961 to a staggering 41,604 tons in 2011, an almost sixfold increase; and the productivity per colony has doubled (Figure 10.3). However, this grand economic success had the serious downside of the defensive behavior of the African bees.

What Kerr had also not foreseen was that the African honey bees would be highly invasive and rapidly spread throughout the American tropics. The looming approach of the killer bees as they moved northward not only caused public panic but also resulted in significant funds being provided for honey bee research aimed at understanding the biology of the "killer" bees and eventually allowing for the development of a means to stop the advance. Honey bee researchers worldwide embarked on the study of these bees and indeed produced exciting results, leading to a profound understanding of the invasive process. It developed into a research field of its own, and searching the Web

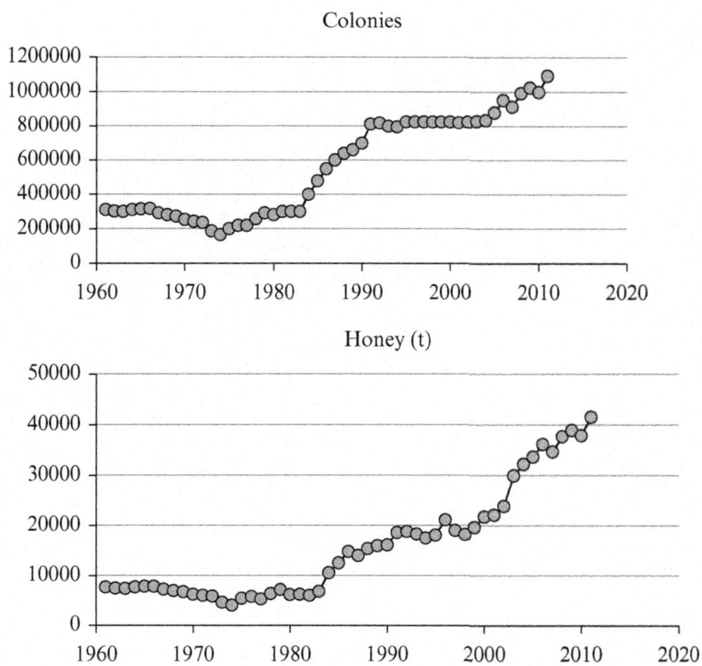

FIGURE 10.3 *Number of managed honey bee colonies (top) and production of honey (bottom) in Brazil after the introduction of African honey bees into the country in the late 1950s.*

of Science for "Africanized honey bee" resulted in the identification of close to 780 publications in peer-reviewed journals (as of July 2015) with more than 12,000 citations in the time span from 1977 to the present. The publications started with Orley Taylor's (1977) rather accurate predictions on past and possible future spread of the bees in the Americas. Despite all the efforts made to halt the spread of the honey bees, they could not be stopped by barriers, traps, or breeding programs (Figure 10.4). African honey bees simply replaced European honey bee colonies. The African honey bees spread as far as they could and were only stopped by natural ecological constraints when they eventually arrived at more temperate climates in the United States in the north and Argentina in the south.

The introduction of African honey bees resulted in a tremendous boost to apiculture, with many satisfied beekeepers quite happy with their introduction after recovering from the initial shock of their arrival (DeJong 1996). The major concern, however, was the impact of the imported honey bees on the native bee

FIGURE 10.4 *The spread of African honey bees in South and North America from their introduction in 1956 until 2015.*

fauna. South and Central America are hotspots for bee diversity, comprising thousands of solitary and social bee species (Michener 2007). In particular, the vast fauna of stingless bee species was expected to be threatened by the huge success of honey bees. This time, it turned out that not just Kerr but also the thousands of native species fauna were lucky. Despite the rapid spread of the African bees, the native bee fauna remained intact. The African honey bees simply replaced European colonies as a consequence of the possession of a set of highly adaptive traits (Spivak 1992). Although the drones of African honey bee colonies out compete European ones (Rinderer et al. 1991), the replacement is primarily driven by swarm takeover as shown by Alice Pinto and collaborators (2004). They showed that replacement was not primarily due to hybridization but, rather, due to simple colony takeover, a behavioral trait characteristic of African honey bees colonies. An absconding swarm takes over a host colony by killing the resident queen and establishing its own queen in the colony (Moritz et al. 2011). As a result, the feral European honey bee population in Texas changed from 0% to 90% African honey bees over a period of six years. From the outside, the beekeeper will not initially notice any difference. The colony is still alive and active. Only after a few weeks, when the first African worker honey bees join the guards at the entrance to the colony, will the beekeeper notice that the colony has changed and, if unprepared, be in for a rather unpleasant experience. It took many more years for bidirectional gene flow to result in an admixed honey bee population of both African and European types in the same population (Rangel et al. 2016).

Despite the negative effects of their defensive behavior, it was exceptionally fortunate that the spread of African honey bees affected only the European honey bee populations and not any other wild bee species. This was of course a major concern for all pollination ecologists studying bees other than the honey bee. David Roubik, who painstakingly studied the competition between *Apis* and non-*Apis* bees on the American continent (Roubik 1978, 1980, 1981, 1983; Roubik and Buchmann 1984), found that there had been no change at all in pollination stability and no species declines in response to the spread of the African honey bees (Roubik 2000, 2001; Roubik and Wolda 2001; Roubik and Villanueva-Gutierrez 2009). Also, the honey bees did not acquire any novel diseases such as those transferred after their introduction to Asia, nor did they spread novel diseases (Roubik and Reyes 1984; Rosenkranz 1999; Teixeira et al. 2008; Calderon et al. 2010; Santos et al. 2014). The Africanized honey bees have even been shown to be more resistant to the parasitic *Varroa* mite (Calderon et al. 2010). Obviously, the niche overlap with native bee species was much less relevant than in Asia, where *A. mellifera* was exposed to other *Apis* species with very similar biology. Thus, although the African honey bees are now an integral part of the wild bee fauna in the American tropics, nothing happened other than the spread of an additional pollinator excellently adapted to tropical habitat conditions—indeed, a most fortunate end to a situation that might have seen the eradication of the entire pollinator community in that region. Today, the honey bees are no longer considered a biodiversity threat in America, and with current concerns about pollinator declines, the opposite has come true. African honey bees have even been considered to be important pollinators for otherwise endangered plant species (Sanguinetti and Singer 2014).

So to evaluate the case in hindsight: With today's knowledge of honey bee biology and the recognized need to conserve biodiversity, it is clear that Warwick Kerr should never have embarked on his experiment. However, only because of his lack of knowledge in the 1950s of the current state of the art in conservation biology do we now know and understand how tropical honey bees outcompeted maladapted honey bee colonies from temperate climates. We know that natural selection is very important with regard to keeping bees. Natural selection will win, not the bee breeder. We should stop thinking that we can easily interfere with a honey bee population that can thrive and prosper in the wild. Despite the various maladaptations highlighted in this book, it is clear that honey bee colonies can survive very efficiently in a huge variety of habitats and under diverse climatic conditions as long as there is sufficient space to provide for sufficient nesting sites and forage.

{ 11 }

A Silver Lining for the Future of Bees?

Ein Ende kommt nicht von selbst. Man muss es schließlich setzen.
The end does not come automatically. Ultimately one has to call a halt.
—Karl von Frisch (1965, p. v)

11.1. Where do honey bees go from here?

If we believed the mass media reports of the past decade, then honey bees would disappear from the earth because humans are destroying the planet. If the latter is true, honey bees will indeed disappear, but if we believed such a scenario, we would not have written this book in the first place because the prospect of having readers would disappear. Humans can clearly destroy the planet, but perhaps they will remain sufficiently sane that there is some value in discussing the future perspective of honey bees while the world is still in place in a meaningful way. We do not want to add to the multitude of dramatic global change scenarios but, rather, keep the perspective focused on honey bees.

Honey bees have survived various ice ages and considerably warmer global climates than we are experiencing today (see Figure 2.5). From the perspective of climate change, the honey bees may actually be in a very comfortable position—much more comfortable than most of the other >20,000 bee species. Honey bees can survive in almost any climatic zone with the exception of an arctic climate. As long as there are sufficient bouts of rich nectar and pollen forage throughout the year, the honey bees will be fine. They can easily survive periods of drought or cold spanning more than five months, which makes them exceptionally robust and reliable pollinators. If they are not able to survive, most, if not all, of the other bee pollinators will have disappeared before they do. Although the quote "If the bee disappears from the surface of the earth, man would have no more than four years to live" was almost certainly not coined by Albert Einstein (Calaprice, 2010, p. 479), it may nevertheless be sadly prophetic. The Belgian writer Maurice Maeterlinck (1901), who received the Nobel Prize for literature in 1911, addressed the issue in *The Life of the Bee* by stating,

> You will probably more than once have seen her fluttering about the bushes, in a deserted corner of your garden, without realising that you were carelessly watching the venerable ancestor to whom we probably owe most of our flowers and fruits (for it is actually estimated that more than a hundred thousand varieties of plants would disappear if the bees did not visit them), and possibly even our civilisation, for in these mysteries all things intertwine. (p. 279)

Indeed, the global economic value of pollination has been estimated to exceed €310 billion annually (Gallai et al. 2009). Whether or not this loss of pollination is going to cause the human race to die out in four years may be questionable. We believe that any hostile environment that precludes the survival of honey bees (e.g., an artic climate) may critically impact most life forms on earth, including humankind. We are actually inclined to disagree with the quote attributed to Einstein and think that if the honey bees disappear from this planet, humans will likely have disappeared well before that point is reached.

However, as long as humans are still around, they are likely to care for honey bee colonies as beekeepers, which gives both species an opportunity to survive. But not all honey bees are treated well. As we have previously discussed, it is the western honey bee *Apis mellifera* that benefits from special treatment by humans, and the other honey bee species may become increasingly at risk due to *mellifera* beekeeping (Moritz and Erler 2016). Some honey bees are even treated as invasive pests. For example, the spread of the dwarf honey bee *Apis florea* into the Middle East has sparked the initiation of eradication programs in Israel and Jordan. But even the western honey bee *A. mellifera* is not safe from persecution. In Australia, conservationists are concerned that feral honey bee colonies might destabilize entire ecosystems by generating an invasion vortex of honey bee-pollinated invasive plants.

Indeed, invasive plants provide a rich mass forage for honey bees, and because they are specialist exploiters of mass nectar flows, they will inevitably pollinate the invaders. The most prominent example may be Paterson's curse (*Echium plantagineum*). It is a herb with a lovely violet flower originally imported to Australia as an ornamental that has now invaded vast areas of the continent (Figure 11.1). Although it is a good forage plant for honey bee colonies, it is toxic to vertebrates, including pasture species. Because it is also herbicide resistant, control measures appear to be difficult.

11.1.1. HONEY BEE ANGST DOWN UNDER

One control solution was to eliminate honey bees because they were considered to be the main pollinator and hence a co-culprit. However, the evidence is weak at best that other pollinators might not take advantage were the honey bees to be removed. Just showing that honey bees are abundant is not sufficient to exclude

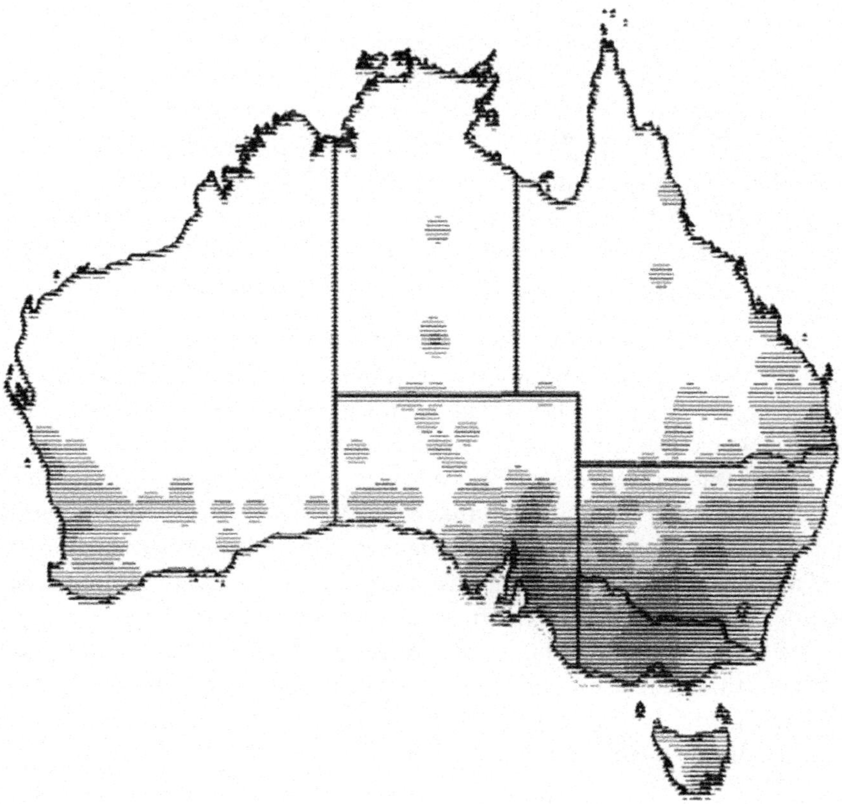

FIGURE 11.1 *Current distribution of Paterson's curse (Echium plantagineum) in Australia. Grayscales indicate different levels of abundance.*

the possibility that other pollinators would fill the pollinator gap for invasives. Thus, much to the dismay of beekeepers, honey bees have been the target of large-scale eradication programs using bait trapping of feral colonies. All these programs seem to have had limited impact at best, and the bees have continued to thrive, which demonstrates the vigor of honey bee populations in the wild.

Even more concerns emerged with the arrival of the Asian honey bee *Apis cerana* in Australia in 2007. The Australian government reacted with an action plan

Including

- an eradication programme to detect and destroy any Asian honey bee swarms and nests
- the introduction of movement restrictions controlling managed bees and beekeeping equipment
- a notification system so all hives reported could be destroyed.

> Four years later and with accumulating experience "Biosecurity Queensland" managed the response to the detection of the Asian honey bee in Cairns on behalf of the Asian Honey Bee National Management Group (NMG). The NMG comprises members from national and state/territory agricultural departments, AHBIC and Plant Health Australia met in January 2011 to conclude that the eradication of the Asian honey bee was no longer technically feasible. A comprehensive manual by Rice et al (2016) addresses the issue rescaling the efforts to surveillance and monitoring. . . .

Although the chance to achieve eradication may have been lost in Australia, and the role of honey bees with regard to invasive plants is still a controversial issue, it is clear that two species of honey bee will be an integral part of Australian biodiversity for the foreseeable future.

The extraordinary adaptability of *A. mellifera* colonies may be a major biodiversity issue resulting from the tremendous growth of beekeeping using this species in Asia. Use of *A. mellifera* in Asia for honey production may indeed become the main challenge for the genus *Apis* in the future, and we encourage bee researchers to carefully monitor interactions between imported *A. mellifera* and the other honey bee species in Asia (Pirk, Crewe, and Moritz 2017) so that there is no catastrophic loss of biodiversity such as occurred in North America during the Miocene (see Figure 2.4; Engel, Hinijosa-Diaz, and Rasnitzyn 2009).

11.2. Emerging into the light

After having dealt with all the dark sides, mishaps, and maladaptations, we must now conclude by asking another question: Why do honey bees still exist given the many maladaptive blunders they make and the amount of conflict and competition in the colony? How do they possibly get by?

The answer is actually rather simple: It depends on all the marvellous stories told in all the other monographs on the honey bee by others. Over many centuries, over and over again honey bee scientists have admired the achievements of the colony and the individual bees. Today, we believe the success of the honey bee colony is driven by the bees' skill at exploiting mass forage. They do not need a few special flowers—they need a flowering tree. They do not need a small strip of flowers alongside the field—they need the entire field. As soon as foraging becomes a numbers game, the honey bee is in its element. It is then that they outcompete other pollinators by combining their individual skills that result from learning, memory, and language with the power of the colony through self-organized recruitment of large numbers of workers to exploit mass forage. Honey bee colonies generally do not exploit rare floral sources. Although they are often considered generalist pollinators, they are actually not. They are specialists in the very true sense—specializing on mass forage flora. In this way, they provide themselves with the honey stores required to bridge the periods of dearth. If forage is patchy, they are regularly less efficient than many other bee species that are very often even more abundant than the honey bees unless apiaries are in the vicinity.

The other major selective advantage is colonial homeostasis. Honey bees can survive when others simply freeze or die out. They can buffer harsh environmental conditions and are highly resilient, even in relation to the artificial conditions imposed by beekeepers, including frames to hold their combs, boxes made from all sorts of materials, queen excluder grids, or other arrangements that are not encountered in the wild.

11.3. Where does apiculture go from here?

Throughout the history of beekeeping, humans have manipulated bee colonies from all possible sides—from the top, the bottom, the front, the back, and either side of the container provided to the colony. The extreme resilience of honey bees makes this diversity of apicultural approaches possible.

Apicultural exploitation of honey bees has resulted in the appearance of massive beekeeping operations in which thousands of colonies are handled on an industrial scale. In extensive commercial beekeeping, there is no time to search for the queen or check her egg-laying capacity. Monitoring each and every colony with regard to brood development, colony behavior, honey harvest, or colony health will inevitably be neglected as the colony numbers in an operation grow. At the same time, the high density of colonies in large commercial operations with several thousand colonies exacerbates disease problems through enhanced pathogen transmission and selection for more virulent pathogens (for another cause of colony loss, see Figure 11.2). Apiculture might benefit from harnessing natural selection and using locally adapted honey bees rather than pursuing the chimera of one jack-of-all-trades bee for all purposes. Currently, the divergence between hobby and commercial beekeepers seems to be increasing in relation to managed colony numbers and intensity of management. Whereas city beekeeping with one to three colonies is increasing in Western societies, the iconic rural smallholder beekeeper seems to be disappearing. Apiculture seems to be polarizing into operations with either too few or too many colonies to be sustainable.

FIGURE 11.2 *Another form of colony loss in large migratory beekeeping operations.*

11.4. Where does honey bee research go from here?

After having now completed a full volume on open questions, we would of course like to see some, if not all, of the "maladaptions" we have listed proven to be wrong. The main riddle, however, remains: Why does the colony not collapse in a complexity catastrophe? Although the integration of individual bees in the colony is well understood and the basic ground plan is described in Robert Page's *Spirit of the Hive* (2013), which beautifully describes the principles of colony organization, the full delineation of colony-level responses remains hidden in Karl von Frisch's magic well, with researchers continuing to draw new knowledge from the well and hence also generating many additional questions. Understanding this complexity may require not just the quantification of a few target individuals in the colony but also the assessment of all individuals, including the analysis and understanding of all interactions. So the challenges currently faced in the analysis of interactions among thousands of genes in genomes to gain a comprehensive understanding of the construction of a phenotype will be magnified at the colony level, where we need to understand the interactions of thousands of genotype-driven phenotypes that eventually make up the colony. Approaches that fail to appropriately acknowledge the individuality of each and every bee contributing to colony success are likely to fail to reach this goal. We are convinced that the growth of computational power in combination with novel and more accessible techniques that allow for the analysis of complex biological systems from the genomic level to complex social webs, will provide the essential elements that will make such detailed analyses possible.

Karl von Frisch's magic well of honey bee research turns out to be even deeper than we imagined. It is clearly too deep for the two authors of this book, who remain puzzled about many facets of honey bee biology despite long professional careers trying to understand how honey bees operate. In the end, we rejoin Karl von Frisch to watch the magic well refilling faster and faster the more that is drawn from it. Yet wells are pitch dark at the bottom, and bottomless wells are even darker. So do not be surprised to find more dark sides to many of the conundrums yet to be explored.

{ REFERENCES }

Abrol DP (2013). *Asiatic Honeybee* Apis cerana*: Biodiversity Conservation and Agricultural Production*. Springer Verlag, Dordrecht, the Netherlands.

Adam B (1983). *In Search of the Best Strains of Bees*. Northern Bee Books, Hebden Bridge, United Kingdom.

Aizen MA, Harder LD (2009). The global stock of domesticated honey bees is growing slower than agricultural demand for pollination. *Curr Biol* 19: 915–918.

Allen MF, Ball BV, Underwood BA (1990). An isolation of *Melissococcus pluton* from *Apis laboriosa. J Invertebrate Pathol* 55: 439–440.

Allsopp MH (1993). Summarized overview of the Capensis Problem. *S Afric Bee J* 65: 127–136.

Allsopp MH, Crewe R (1993). The Cape honeybee as a Trojan horse rather than the hordes of Jenghiz Khan. *Am Bee J* 133: 121–123.

Anderson C (2001). The adaptive value of inactive foragers and the scout–recruit system in honey bee (*Apis mellifera*) colonies. *Behav Ecol* 12: 111–119.

Anderson DL (2010). *Control of Asian Honeybees in the Solomon Islands*. Australian Centre for International Agricultural Research, Canberra, Australia.

Anderson DL, Annand N, Lacey M, Ete S (2012). *Control of Asian Honey Bees in Solomon Islands*. Australian Centre for International Agricultural Research, Canberra, Australia.

Anderson RH (1963). The laying worker in the Cape Honeybee, *Apis mellifera capensis. J apic Res* 2: 85–92.

Aristotle (350 BCE). *The History of Animals*, translated by D'Arcy Wenthworth Thompson. Book V, part 21, The Internet Classics Archive by Daniel C. Stevenson, Web Atomics.

Arnold G, Quenet B, Masson C (2000). Influence of social environment on genetically based subfamily signature in the honey bee. *J Chem Ecol* 26: 2321–2333.

Asencot M, Lensky Y (1976). The effect of sugars and juvenile hormone on the differentiation of the female honey bee larvae (*Apis mellifera* L.) to queens. *Life Sci* 18: 693–699.

Asencot M, Lensky Y (1984). Juvenile hormone induction of "queenliness" on female honey bee (*Apis mellifera* L.) larvae reared on worker jelly and on stored royal jelly. *Comp Biochem Physiol B* 78: 109–117.

Asencot M, Lensky Y (1985). The phagostimulatory effect of sugars on the induction of "queenliness" in female honey bee (*Apis mellifera* L.) larvae. *Comp Biochem Physiol A* 81: 203–208.

Asencot M, Lensky Y (1988). The effect of soluble sugars in stored royal jelly on the differentiation of female honey bee (*Apis mellifera* L.) larvae to queens. *Insect Biochem* 18: 127–133.

Ayasse M, Paxton RJ, Tengö J (2001). Mating behavior and chemical communication in the order Hymenoptera. *Annu Rev Entomol* 46: 31–78.

Barth MB, Moritz RFA, Walter C, Pirk W, Kraus FB (2013). Male-biased dispersal promotes large scale gene flow in a subterranean army ant, *Dorylus (Typhlopone) fulvus. Popul Ecol* 55: 523–533.

Bauer D, Bienefeld K (2013). Hexagonal comb cells of honey bees are not produced via a liquid equilibrium process. *Naturwissenschaften* 100: 45–49.

Beaurepaire AL, Truong TA, Fajardo AC, Dinh TQ, Cervancia C, Moritz RFA (2015). Host Specificity in the Honeybee Parasitic Mite, *Varroa spp. Apis mellifera* and *Apis cerana. PLoS One* 10(8): e0135103.

Beekman M, Calis JNM, Boot WJ (2000). Insect behaviour—Parasitic honeybees get royal treatment. *Nature* 404: 723.

Beekman M, Fathke RL, Seeley TD (2006). How does an informed minority of scouts guide a honeybee swarm as it flies to its new home? *Anim Behav* 71: 161–171.

Beggs JR, Brockerhoff EG, Corley JC, Kenis M, Masciocchi M, Muller F, Rome Q, Villemant, C (2011). Ecological effects and management of invasive alien *Vespidae. Biocontrol* 56: 505–526.

Behrens D, Hunag Q, Gessner C, Rosenkranz P, Frey E, Locke B, Moritz RFA, Kraus FB (2011). Three QTL in the honey bee *Apis mellifera* L. suppress reproduction of the parasitic mite *Varroa destructor. Ecol Evol* 1: 4. doi:10.1002/ece3.17.

Berg S, Koeniger N, Koeniger G, Fuchs S (1997). Body size and reproductive success of drones (*Apis mellifera* L.). *Apidologie* 28: 449–460.

Betts AD (1939) The mating flight: a summary of present knowledge. *Bee World* 20: 20–24, 33–36.

Beyaert L, Greggers U, Menzel R (2012). Honey bees consolidate navigation memory during sleep. *J Exp Biol* 215: 3981–3988.

Beye M, Hasselmann M, Fondrk MK, Page RE, Omholt SW (2003). The gene csd is the primary signal for sexual development in the honeybee and encodes an SR-type protein. *Cell* 114: 419–429.

Biesmeijer J, Seeley TD (2005). The use of waggle dance information by honey bees throughout their foraging careers. *Behav Ecol Sociobiol* 59: 133–142.

Biganski S, Kurze C, Müller MY, Moritz RFA (2018). Social response of healthy honeybees towards Nosema ceranae-infected workers: care or kill? *Apidologie.* https://doi.org/10.1007/s13592-017-0557-8

Bloch G (2010). The social clock of the honey bee. *J Biol Rhythms* 25: 307–317.

Bloch G, Rubinstein CD, Robinson GE (2004). Period expression in the honey bee brain is developmentally regulated and not affected by light, flight experience, or colony type. *Insect Biochem Mol Biol* 34: 879–891.

Bloch G, Shemesh Y, Robinson GE (2006). Seasonal and task-related variation in free running activity rhythms in honey bees (*Apis mellifera*). *Insectes Soc* 53: 115–118.

Boes KE (2010). Honey bee colony drone production and maintenance in accordance with environmental factors: An interplay of queen and worker decisions. *Insect Soc* 57: 1–9.

Boomsma JJ, d'Ettorre P (2013). Nice to kin and nasty to non-kin: Revisiting Hamilton's early insights on eusociality. *Biol Lett* 9: 20130444.

Boot WJ, Calis JNM, Beetsma J, Hai DM, Lan NK, Van Tran T, Trung LQ, Minh NH (1999). Natural selection of *Varroa jacobsoni* explains the different reproductive strategies in colonies of *Apis cerana* and *Apis mellifera. Exp Appl Acarol* 23: 133–144.

Bratkowski J, Pirk CWW, Neumann P, Wilde J (2012). Genotypic diversity in queenless honey bee colonies reduces fitness. *J Apic Res* 51: 336–341.

Brockmann A, Brückner D, Crewe RM (1998). The EAG response spectra of workers and drones to queen honeybee mandibular gland components: The evolution of a social signal. *Naturwissenschaften* 85: 283–285.

Brown MJF, Schmid-Hempel P (2003). The evolution of female multiple mating in social hymenoptera. *Evolution* 57: 2067–2081.

Büchler R, Berg S, Le Conte Y (2010). Breeding for resistance to *Varroa destructor* in Europe. *Apidologie* 41: 393–408.

Buchwald R, Breed MD, Greenberg AR (2008). The thermal properties of beeswaxes: Unexpected findings. *J Exp Biol* 211: 121–127.

Butenandt A, Beckmann R, Stamm D, Hecker E (1959). Über den Sexuallockstoff des Seidenspinners *Bombyx mori*—Reindarstellung und Konstitution. *Z Naturforsch B* 14: 283–284.

Butler C (1609). *The Feminine Monarchie*. Joseph Barnes, Oxford, UK.

Butler CG, Callow RK, Chapman JR (1964). 9-Hydroxydec-trans-2-enoic acid pheromone stabilizing honeybee swarms. *Nature* 201: 733.

Butler CG, Simpson J (1967). Pheromones of queen honeybee (*Apis mellifera* L.) which enable her workers to follow her when swarming. *Proc R Entomol Soc Lond Ser A* 42: 149–154.

Buttstedt A, Ihling CH, Pietzsch M, Moritz RFA (2016). Royalactin is not a royal making of a queen. *Nature* 537: 10–12.

Buttstedt A, Moritz RFA, Erler S (2014). Origin and function of the major royal jelly proteins of the honeybee (*Apis mellifera*) as members of the yellow gene family. *Biol Rev* 89: 255–269.

Calaprice A (2010). *The Ultimate Quotable Einstein*. Princeton University Press, Princeton, NJ.

Calderon RA, van Veen JW, Sommeijer MJ, Sanchez LA (2010). Reproductive biology of *Varroa destructor* in Africanized honey bees (*Apis mellifera*). *Exp Appl Acarol* 50: 281–297.

Camargo JMF, Roubik DW (1991). Systematics and bionomics of the apoid obligate necrophages: The *Trigona hypogea* group (Hymenoptera: Apidae, Meliponinae). *J Linnean Soc* 44: 13–39.

Camazine S (1993). The regulation of pollen foraging by honey-bees—How foragers assess the colony need for pollen. *Behav Ecol Sociobiol* 32: 265–272.

Camazine S, Sneyd J (1991). A model of collective nectar source selection by honey-bees—Self-organization through simple rules. *J Theor Biol* 149: 547–571.

Cardinal S, Danforth BN (2011). The antiquity and evolutionary history of social behavior in bees. *PLoS One* 6(6): e21086. doi:10.1371/journal.pone.0021086

Cardinal S, Danforth BN (2013). Bees diversified in the age of eudicots. *Proc R Soc Ser B* 280: 20122686. doi:10.1098/rspb.2012.2686

Chippindale AK, Gibson JR, Rice WR (2001). Negative genetic correlation for adult fitness between sexes reveals ontogenetic conflict in *Drosophila*. *Proc Natl Acad Sci USA* 98: 1671–1675.

Cobey S, Schley P (2002). Innovations in instrumental insemination—The compact, versatile right and left handed Schley Model II instrument. *Am Bee J* 142: 433–435.

Coelho JR (1991). The effect of thorax temperature on force production during tethered flight in honeybee (*Apis mellifera*) drones, workers, and queens. *Physiol Zool* 64: 823–835.

Coelho JR (1996). The flight characteristics of drones in relation to mating. *Bee Sci* 4: 21–25.

Cole BJ, Wiernasz DC (1999). The selective advantage of low relatedness. *Science* 285: 891–893.

Costa C, Lodesani M, Bienefeld K (2012). Differences in colony phenotypes across different origins and locations: Evidence for genotype by environment interactions in the Italian honeybee (*Apis mellifera ligustica*)? *Apidologie* 43: 634–642.

Cotter SC, Kilner RM (2010). Personal immunity versus social immunity. *Behav Ecol* 21: 663–668.

Couto A, Monceau K, Bonnard O, Thiéry D, Sandoz J-C (2014). Olfactory attraction of the hornet *Vespa velutina* to honeybee colony odors and pheromones. *PLoS One* 9(12): e115943. http://dx.doi.org/10.1371/journal.pone.0115943

Couvillon MJ, Hughes WHO, Perez-Sato JA, Martin SJ, Roy GGF, Ratnieks FLW (2010). Sexual selection in honey bees: Colony variation and the importance of size in male mating success. *Behav Ecol* 21: 520–525.

Couvillon MJ, Segers FHID, Cooper-Bowman, R, Truslove G, Nascimento DL, Nascimento FS, Ratnieks, FLW (2013). Context affects nestmate recognition errors in honey bees and stingless bees. *J Exp Biol* 216: 3055–3061.

Cox-Foster DL, Conlan S, Holmes EC, Palacios G, Evans JD, Moran NA, Quan PL, Briese T, Hornig M, Geiser DM, et al. (2007). A metagenomic survey of microbes in honey bee colony collapse disorder. *Science* 318: 283–286.

Crailsheim K (1991). Interadult feeding of jelly in honeybee (*Apis-mellifera* L.) colonies. *J Comp Physiol B* 161: 55–60.

Crailsheim K (1992). The flow of jelly within a honey bee colony. *J Comp Physiol B* 162: 681–689.

Crailsheim K (1998). Trophallactic interactions in the adult honeybee (*Apis mellifera* L.). *Apidologie* 29: 97–112.

Crailsheim K, Hrassnigg N, Stabentheiner A (1996). Diurnal behavioural differences in forager and nurse honey bees (*Apis mellifera carnica* Pollm). *Apidologie* 27: 235–244.

Crane E (1975). *Honey: A Comprehensive Survey*. Heinemann, London.

Crane E (1983). *The Archaeology of Bee-Keeping*. Duckworth, London.

Cremer S, Armitage SA, Schmid-Hempel P (2007). Social immunity. *Curr Biol* 17: R693–R702.

Cronin AL, Molet M, Doums C, Monnin T, Peeters C (2013). Recurrent evolution of dependent colony foundation across eusocial insects. *Ann Rev Entomol* 58: 37–55.

Crozier RH (1981). *Animal Cytogenetics 3, Insecta 7, Hymenoptera*. Gebrüder Borntraeger, Berlin. (Original work published 1975)

Crozier RH (1971). Heterozygosity and sex determination in haplo-diploidy. *Am Nat* 105: 399–412.

Dampney JR, Barron AB, Oldroyd BP (2002). Policing of adult honey bees with activated ovaries is error prone. *Insect Soc* 49: 270–274.

Darchen R (1956). La reine d'*Apis-mellifica* et les constructions cirieres. *Compt Rend Hebd Seances Acad Sci* 243: 439–441.

Darwin C (1859). *On the Origin of Species by Means of Natural Selection*. John Murray, London.

Darwin C (1890). *The Descent of Man and Selection in Relation to Sex*, 2nd ed. John Murray, London.

De Boer JG, Groenen MAM, Pannebakker BA, Beukeboom LW Kraus RHS (2015). Population-level consequences of complementary sex determination in a solitary parasitoid. *BMC Evol Biol* 15: 98. doi:10.1186/s12862-015-0340-2.

De la Rua P, Jaffe R, Dall'Olio R, Munoz I, Serrano J (2009). Biodiversity, conservation and current threats to European honeybees. *Apidologie* 40: 263–284.

De Meulemeester T, Engel MS, Michez D (2015). *Checklist of the Bee Fossils (Hymenopter a: Apoidea: Anthophila)*. Retrieved from http://beefossils.myspecies.info

de Reaumur M (1750). *Memoires pour Server a l'histoire des Insects*. De L'imprimerie royale, Paris.

Debevec AH, Cardinal S, Danforth BN (2012). Identifying the sister group to the bees: A molecular phylogeny of *Aculeata* with an emphasis on the superfamily Apoidea. *Zool Scripta* 41: 527–535.

Degen J, Kirbach A, Reiter L, Lehmann K, Norton P, Storms M, Koblofsky M, Winter S, Georgieva PB, Nguyen H, Chamkhi H, Meyer H, Singh PK, Manz G, Greggers U Menzel R (2016). Honeybees Learn Landscape Features during Exploratory Orientation Flights. *Curr Biol* 26: 2800–2804.

DeJong, D (1996). Africanized honey bees in Brazil, forty years of adaptation and success. *Bee World* 77: 67–70.

de Villers JH (1883). Cape Bees and "Animal Intelligence." *Nature* 28: 5–6.

Deyel MJ (1980). Supersedure of queens. In Root AI (ed.), *The ABC and XYZ of Bee Culture*, p. 603. AI Root Company, Medina, OH.

Die Bienen—Tödliche Bedrohung, directed by Michael Karen. Wasabi Film, Gruenwald, Germany (2008).

Dobbs A (1750). Concerning bees, and their method of gathering wax and honey. *Philos Trans R Soc* 46: 536–549.

Doner LW (1977). The sugars of honey—A review. *J Sci Food Agric* 28: 443–456.

Dornhaus A, Klugl F, Oechslein C, Puppe F, Chittka L (2006). Benefits of recruitment in honey bees: Effects of ecology and colony size in an individual-based model. *Behav Ecol* 17: 336–344.

Downs SG, Ratnieks FLW (1999). Adaptive shifts in honey bee (*Apis mellifera* L.) guarding behavior support predictions of the acceptance threshold model. *Behav Ecol* 11: 326–333.

Dyer AG, Dorin A, Reinhardt V, Garcia JE, Rosa MGP (2014). Bee reverse-learning behavior and intra-colony differences: Simulations based on behavioral experiments reveal benefits of diversity. *Ecol Modelling* 277: 119–131.

Eban-Rothschild AD, Bloch G (2008). Differences in the sleep architecture of forager and young honeybees (*Apis mellifera*). *J Exp Biol* 211: 2408–2416.

Ellis JD, Pirk CWW, Hepburn HR, Kastberger G, Elzen P (2002). Small hive beetles survive in honey bee prisons by behavioural mimicry. *Naturwissenschaften* 89: 326–328.

Engel MS (2000). A new interpretation of the oldest fossil bee (Hymenoptera: Apidae). *Am Museum Novit* 3296: 1–11.

Engel MS, Hinojosa-Diaz, IA, Rasnitzyn, AP (2009). A honey bee from the Miocene of Nevada and the biogeography of *Apis* (Hymenopteras: Apidae: Apini). *Proc Calif Acad Sci* 60: 23–38.

Erler S, Denner A, Bobis O, Forsgren E, Moritz RFA (2014). Diversity of honey stores and their impact on pathogenic bacteria of the honey bee, *Apis mellifera. Ecol Evol* 4: 3960–3967.

Erler S, Moritz RFA (2016). Pharmacophagy and pharmacophory: Mechanisms of self-medication and disease prevention in the honeybee colony (*Apis mellifera*). *Apidologie* 47(3): 389–411. doi:10.1007/s13592-015-0400-z

Esch H (1967). The sounds produced by swarming honey bees. *Z Vergl Physiol* 56: 408–411.

Es'kov EK, Es'kova MD (2013). Factors influencing wing size and body weight variation in the western honeybee. *Russ J Ecol* 44: 433–438.

Esslen J, Kaissling KE (1976). Zahl und Verteilung antennaler Sensillen bei der Honigbiene (*Apis mellifera* L.). *Zoomorphol* 83: 227–251.

Estoup A, Solignac M, Cornuet JM (1994). Precise assessment of the number of patrilines and genetic relatedness in honeybee colonies. *Proc R Soc B Biol Sci* 258: 1–7.

Evans JD, Aronstein K, Chen YP, Hetru C, Imler J-L, Jiang H, Kanost M, Thompson G.J., Zou Z, Hultmark D (2006). Immune pathways and defence mechanisms in honey bees *Apis mellifera. Insect Mol Biol* 15: 645–656.

Evans JD, Spivak M (2010). Socialized medicine: Individual and communal disease barriers in honey bees. *J Invertebr Pathol* 103: S62–S72.

Fabré JH (2002). *The Life of the Caterpillar*, Honolulu, Hawaii: University Press of the Pacific.

Fabré JH (2002). *Souvenirs Entomologiques*. BiblioLife, Charleston, SC. (Original work publsihed 1879–1907).

Field J (2005). The evolution of progressive provisioning. *Behav Ecol* 16: 770–778.

Fletcher DJC (1977). Evaluation of introductions of European honeybees into eastern and southern Africa. *Proc. 8th Internat Cong IUSSI*, Wageningen, pp. 146–147.

Fletcher DJC (1978). The African bee, *Apis mellifera adansonii*, in Africa. *Ann Rev Entomol* 23: 151–171.

Flottum K (2011). The bicentenary of Revd Lorenzo Lorraine Langstroth: Part 2. *Bee Craft* 24.

Forsgren E, Shi W, Ding G, Liu, ZG, Tran TV, Tang PT, Truong TA, Dinh, TQ, Fries I (2015). Preliminary observations on possible pathogen spill-over from *Apis mellifera* to *Apis cerana. Apidologie* 46: 265–275.

Fouks B, Lattorff HMG (2011). Recognition and avoidance of contaminated flowers by foraging bumblebees (*Bombus terrestris*). *PLoS One* 6: e26328.

Free JB (1957). The food of adult drone honeybees (*Apis mellifera*). *Brit J Anim Behav* 5: 7–11.

Fries I, Bommarco (2007). Possible host-parasite adaptations in honey bees infested by *Varroa destructor* mites. *Apidologie* 38: 525–533.

Fries I, Camazine S (2001). Implications of horizontal and vertical pathogen transmission for honey bee epidemiology. *Apidologie* 32: 199–214.

Fries I, Hansen H, Imdorf A, Rosenkranz P (2003). Swarming in honey bees (*Apis mellifera*) and *Varroa destructor* population development in Sweden. *Apidologie* 34: 389–397.

Fries I, Imdorf A, Rosenkranz P (2006). Survival of mite infested (*Varroa destructor*) honey bee (*Apis mellifera*) colonies in a Nordic climate. *Apidologie* 37: 564–570.

Fu Q, Posth C, Hajdinjak M, Petr M, Mallick S, Fernandes D, Furtwängler A, Haak W, Meyer M, Mittnik A, et al. (2016). The genetic history of Ice Age Europe. *Nature* 534: 200–205.

Fuchs S, Moritz RFA (1999). Evolution of extreme polyandry in the honeybee *Apis mellifera* L. *Behav Ecol Sociobiol* 45: 269–275.

Fuchs S, Schade V (1994). Lower performance in honeybee colonies of uniform paternity. *Apidologie* 25: 155–168.

Fujiwara S, Imai J, Fujiwara M, Yaeshima T, Kawashima T, Kobayashi K (1990). A potent antibacterial protein in royal jelly: Purification and determination of the primary structure of royalisin. *J Biol Chem* 265: 11333–11337.

Gallai N, Salles JM, Settele J, Vassiere BE (2009). Economic valuation of the vulnerability of world agriculture confronted with pollinator decline. *Ecol Econ* 68: 810–821.

Gary NE (1963). Observations of mating behaviour in the honeybee. *J Apic Res* 2: 3–13.

Gherman BI, Denner A, Bobiş O, Dezmirean DS, Mărghitaş LA, Schlüns H, Moritz RFA, Erler S (2014). Pathogen-associated self-medication behavior in the honeybee *Apis mellifera. Behav Ecol Sociobiol* 68: 1777–1784.

Gilliam M, Prest DB, Lorenz BJ (1989). Microbiology of pollen and bee bread: Taxonomy and enzymology of molds. *Apidologie* 20: 53–68.

Goetze G (1940). *Die beste Biene*. Lidehoff, Loth and Michaelis, Leipzig.

Goetze, G (1959). Die Bedeutung des Flügelgeäders für die züchterische Bedeutung der Honigbiene. *Zeitschrift für Bienenforschung* 4: 141–148.

Goetze G (1964). *Die Honigbiene in natürlicher und künstlicher Zuchtauslese*. Parey Verl., Hamburg.

Goins A, Schneider SS (2013). Drone "quality" and caste interactions in the honey bee, *Apis mellifera* L. *Insect Soc* 60: 453–461.

Gołębiowski M, Cerkowniak M, Boguś MI, Włóka E, Dawgul M, Kamysz W, Stepnowski P (2013). Free fatty acids in the cuticular and internal lipids of *Calliphora vomitoria* and their antimicrobial activity. *J Insect Physiol* 59: 416–429.

Gould SJ (1981). Evolution as fact and theory. *Discover* 2: 34–37.

Greco MK, Hoffmann D, Dollin A, Duncan M, Spooner-Hart R, Neumann, P (2010). The alternative Pharaoh approach: Stingless bees mummify beetle parasites alive. *Naturwissenschaften* 97: 319–323.

Hamilton WD (1964). The genetical evolution of social behaviour: I and II. *J Theor Biol* 7: 1–52.

Hamilton WD (1972). Altruism and related phenomeny, mainly in social insects. *Ann Rev Ecol Syst* 3:193–232

Han F, Wallberg A, Webster MT (2012). From where did the Western honey bee (*Apis mellifera*) originate? *Ecol Evol* 2: 1949–1957.

Harari YN (2014). *Sapiens: A Brief History of Humankind*. Vintage, London.

Harbo JR, Harris JW (2009). Responses to *Varroa* by honey bees with different levels of *Varroa* Sensitive Hygiene. *J Apic Res* 48: 156–161.

Hartfield M, Keigthley PD (2012). Current hypotheses for the evolution of sex and recombination. *Integr Zool* 7: 192–209.

Haydak M (1970). Honey bee nutrition. *Annu Rev Entomol* 15: 143–156.

Helantera H, d'Ettorre P (2015). A comparative study of egg recognition signature mixtures in *Formica* ants. *Evolution* 69: 520–529.

Hepburn HR (1986). *Honeybees and Wax—An Experimental Natural History*. Springer-Verlag, Berlin.

Hepburn HR (1995). G.W. Onions—Discoverer of the Capensis trait female eggs from laying workers. *SA Bee J* 67: 5–9.

Hepburn HR, Pirk CWW, Duangphakdee O (2014). *Honeybee Nests: Composition, Structure, Function*. Springer Verlag, Berlin.

Hepburn HR, Radloff SE (1998). *Honeybees of Africa*. Springer-Verlag, Berlin.

Hepburn HR, Radloff SE (2011). *Honeybees of Asia*. Springer-Verlag, Berlin.

Herbert EW (1992). Honey bee nutrition. In Graham JM (ed.), *The Hive and the Honey Bee*. Dadant, Hamilton, IL, pp. 197–233.

Hermann HR (1971). Sting autotomy, a defensive mechanism in certain social Hymenoptera. *Insectes Soc* 18: 111–120.

Hillesheim E, Koeniger N, Moritz RFA (1989). Colony performance in honeybees (*Apis-mellifera-capensis Esch*) depends on the proportion of subordinate and dominant workers. *Behav Ecol Sociobiol* 24: 291–296.

Hölldobler B, Bartz SH (1985). Sociobiology of reproduction in ants. In Hölldobler B, Lindauer M (eds.), *Experimental Behavioral Ecology and Sociobiology*. Fischer Verlag, Stuttgart, pp. 237–257.

Hölldobler B, Haskins, CP (1977). Sexual calling behaviour in primitive ants. *Science* 195: 793–794.

Honeybee Genome Sequencing Consortium (2006). Insights into social insects from the genome of the honey bee *Apis mellifera. Nature* 443: 931–949.

Hussaini SA, Bogusch L, Landgraf T, Menzel R (2009). Sleep deprivation affects extinction but not acquisition memory in honey bees. *Learn Mem* 16: 698–705.

Jaffé R, Dietemann V, Allsopp MH, Costa C, Crewe RM, Dall'olio R, de la Rua P, El-Niweiri MAA, Fries I, Kezic N, et al. (2010). Estimating the density of honeybee colonies across their natural range to fill the gap in pollinator decline censuses. *Conserv Biol* 24: 583–593.

Jaffé R, Dietemann V, Crewe RM, Moritz RFA (2009). Temporal variation in the genetic structure of a drone congregation area: An insight into the population dynamics of wild African honeybees (*Apis mellifera scutellata*). *Mol Ecol* 18: 1511–1522.

Jarsoch A, Stolle E, Crewe RM, Moritz RFA (2011). Alternative splicing of a single transcription factor drives selfish reproductive behavior in honeybee workers (*Apis mellifera*). *Proc Natl Acad Sci USA* 108: 15282–15287.

Jay SC (1966). Drifting of honeybees in commercial apiaries. II. Effect of various factors when hives are arranged in rows. *J Apicult Res* 5: 103–112.

Jay SC (1968). Drifting of honey bees in commercial apiaries iv further studies of the effect of apiary layout. *J Apicult Res* 7: 37–44.

Jay SC (1969). The problem of drifting in commercial apiaries. *Am Bee J* 109: 178–179.

Jean RP (2005). Quantifying a rare event: Pollen theft by honey bees from bumble bees and other bees (Apoidea: Apidae, Megachilidae) foraging at flowers. *J Kansas Entomol Soc* 78: 172–175.

Jeanson R, Weidenmüller A (2014). Interindividual variability in social insects—Proximate causes and ultimate consequences. *Biol Rev* 89: 671–687.

Ji R, Xie B, Yang G, Li D (2003). From introduced species to invasive species—A case study on the Italian bee *Apis mellifera* L. *Chinese J Ecol* 2003: 70–73.

Kaatz FH (2006). Measuring the order in ordered porous arrays: can bees outperform humans? *Naturwissenschaften* 93: 374–378.

Kaiser W (1988). Busy bees need rest, too—Behavioral and electromyographical sleep signs in honeybees. *J Comp Physiol A* 163: 565–584.

Kamakura M (2011). Royalactin induces queen differentiation in honeybees. *Nature* 473: 478–483.

Kapheim KM, Pan H, Li C, Salzberg SL, Puiu D, Magoc T, Robertson HM, Hudson ME, Venkat A, Fischman BJ, et al. (2015). Genomic signatures of evolutionary transitions from solitary to group living. *Science* 348: 1139–1143.

Karihaloo B, Zhang K, Wang J (2013). Honey bee combs: How the circular cells transform into rounded hexagons. *J R Soc Interface* 10(86): 20130299. doi:10.1098/rsif.2013.0299

Kerr WE (1967). The history of the introduction of African bees to Brazil. *S Afr Bee J* 2: 3–5.

Kerstiens G (1996). *Plant Cuticle: An Integrated Functional Approach.* BIOS, Oxford, UK.

Kidner J, Moritz RFA (2016). Conditions for the invasion of male-haploidy in diploid populations. *J Evol Biol* 29: 1804–1811.

Killer Bees, directed by Curtis Harrington. R.S.O. Films, London (1974).

Killer Bees! directed by Penelope Buitenhuis. Shavick Entertainment, Victoria, British Columbia, Canada (2002).

Klein BA, Klein A, Wray MK, Mueller UG, Seeley TD (2010). Sleep deprivation impairs precision of waggle dance signaling in honey bees. *Proc Natl Acad Sci USA* 107: 22705–22709.

Klein BA, Olzsowy KM, Klein A, Saunders KM, Seeley TD (2008). Caste-dependent sleep of worker honey bees. *J Exp Biol* 211: 3028–3040.

Klein BA, Seeley TD (2011). Work or sleep? Honey bee foragers opportunistically nap during the day when forage is not available. *Anim Behav* 82: 77–83.

Klein BA, Stiegler M, Klein A, Tautz J (2014). Mapping sleeping bees within their nest: Spatial and temporal analysis of worker honey bee sleep. *PLoS One* 9: e102316.

Kleinhenz M, Bujok B, Fuchs S, Tautz H (2003). Hot bees in empty broodnest cells: heating from within. *J Exp Biol* 4217–4231.

Kocher SD, Grozinger C (2011). Cooperation, conflict and the evolution of queen pheromones. *J Chem Ecol* 37: 1263–1275.

Koeniger G, Koeniger N, Ellis J, Connor L (2014). *Mating biology of honey bees (*Apis mellifera*).* Wicwas Press, Kalamazoo, MI.

Koetz AH (2013a). Ecology, behaviour and control of *Apis cerana* with a focus on relevance to the Australian incursion. *Insects* 4: 558–592.

Koetz AH (2013b). *Spread of* Apis cerana *in Australia, 2007–2012.* Asian Honey Bee Transition to Management Program, Department of Agriculture, Fisheries and Forestry, Queensland, Australia.

Kotthoff U, Wappler T, Engel M (2013). Greater past disparity and diversity hints at ancient migrations of European honey bee lineages into Africa and Asia. *J Biogeogr* 40: 1832–1838.

Kraus FB, Gerecke E, Moritz RFA (2011). Shift work has a genetic basis in honeybee pollen foragers (*Apis mellifera* L.). *Behav Genet* 41: 323–328.

Kraus FB, Neumann P, Moritz RFA (2005). Genetic variance of mating frequency in the honeybee (*Apis mellifera* L.). *Insect Soc* 52: 1–5.

Kritsky, G (2010). *The Quest for the Perfect Hive: A History of Innovation in Bee Culture.* Oxford University Press, New York, USA.

Kryger P, Kryger U, Moritz RFA (2000). Genotypical variability for the tasks of water collecting and scenting in a honey bee colony. *Ethology* 106: 769–779.

Kryger P, Moritz RFA (1997). Lack of kin recognition in swarming honeybees (*Apis mellifera*). *Behav Ecol Sociobiol* 40: 271–276.

Kurze C, Routtu J, Moritz RFA (2016). Parasite resistance and tolerance in honeybees at the individual and social level. *Zoology* 119: 290–297.

Laroca S, Winston ML (1978). Interaction between *Apis mellifera* and *Bombus pennsylvanicus* (Hymenoptera: Apidae) on the flowers of tall thistle: honey bees gather pollen from bodies of bumble-bees. *J Kansas Entomol Soc* 51: 274–275.

Lattorff HMG, Moritz RFA (2016). Context dependent bias in honeybee queen selection: Swarm versus emergency queens. *Behav Ecol Sociobiol* 70: 1411–1417.

Lattorff HMG, Moritz RFA, Crewe RM, Solignac M (2007). Control of reproductive dominance by the thelytoky gene in honeybees. *Biol Lett* 3: 292–295.

Le Conte Y, Hefetz A (2008). Primer pheromones in social Hymenoptera. *Annu Rev Entomol* 53: 523–542.

Le Conte Y, De Vaublanc G, Crauser D, Jeanne F, Rousselle, Becard JM (2007). Honey bee colonies that have survived *Varroa destructor. Apidologie* 38: 566–572.

Le Conte Y, Mohammedi A, Robinson GE (2001). Primer effects of a brood pheromone on worker behavioural development. *Proc R Soc Lond* 268: 163–168.

Leadbeater E, Chittka L (2007). Social learning in insects—From miniature brains to consensus building. *Curr Biol* 17: R703–R713.

Leoncini I, Le Conte Y, Costagliola G, Plettner E, Toth AL, Wang MW, Huang Z, Becard JM Crauser D, Slessor KN, et al. (2004). Regulation of behavioral maturation by a primer pheromone produced by adult worker honey bees. *Proc Nat. Acad Sci USA* 101: 17559–17564.

Lin CL, Chen R, Chen CH, Liu CY (2007). Light enhancement by the formation of an Al oxide honeycomb nanostructure on the n-GaN surface of thin-GaN light-emitting diodes. *Appl Phys Lett* 90: 242106.

Lindauer M (1955). Schwarmbienen auf Wohnungssuche. *Z Vergl Physiol* 37: 263–324.

Lindauer M (1961). *Communication Among Social Bees*. Harvard University Press, Cambridge, MA.

Locke B, Fries I (2011). Characteristics of honey bee colonies (*Apis mellifera*) in Sweden surviving *Varroa destructor* infestation. *Apidologie* 42: 533–542.

Loper GM, Wolf WW, Taylor OR (1992). Honey-bee drone flyways and congregation areas—Radar observations. *J Kansas Entomol Soc* 65: 223–230.

Loper GM, Wolf WW, Taylor OR (1993). Radar detection of drones responding to honeybee queen pheromone. *J Chem Ecol* 19: 1929–1938.

Lopez JCG, Kraus FB (2009). Cherchez la femme? Site choice of drone congregations in the stingless bee *Scaptotrigona mexicana. Anim Behav* 77: 1247–1252.

Lundie AE (1925). The flight activities of the honeybee. *USDA Bull* 1328.

Mackensen O (1951). Viability and sex determination in the honey bee (*Apis mellifera* L.). *Genetics* 36: 500–509.

Maeterlinck M (1901). *The Life of the Bee*. George Allen and Unwin, London.

Martensson L (2010). *Emissions from Volvo's trucks*. Volvo Truck Corporation 01 Reg. No 20640100005: 1–2.

Martin SJ, Kryger P (2002). Reproduction of *Varroa destructor* in South African honeybees: Does cell space influence *Varroa* male survivorship? *Apidologie* 33: 51–61.

Marx K (1845). *The German Ideology. Part I: Feuerbach. Opposition of the Materialist and Idealist Outlook: B. The Illusion of the Epoch*. https://www.marxists.org/archive/marx/works/download/Marx_The_German_Ideology.pdf

Mateus S, Noll FB (2004). Predatory behavior in a necrophagous bee *Trigona hypogea* (Hymenoptera; Apidae, Meliponini). *Naturwissenschaften* 91: 94–96.

Matsuura M, Sakagami SF (1973). A bionomic sketch of the giant hornet, *Vespa mandarinia*, a serious pest for Japanese apiculture. *J Fac Sci Hokkaido Univ Ser VI, Zool* 19: 125–162.

Mattila HR, Seeley TD (2007). Genetic diversity in honey bee colonies enhances productivity and fitness. *Science* 317: 362–364.

Maynard Smith J (1978). *The Evolution of Sex*. Cambridge University Press, Cambridge, UK.

Maynard Smith J (1981). Will a sexual population evolve to an ESS? *Am Nat* 117: 1015–1018.

Maynard Smith J, Price GR (1973). The logic of animal conflict. *Nature* 246: 15–18.

Méndes de Torres L (1586). *Tratado breve de la cultivacion cura de las colmenas*. Author, Alcala de Henares, Spain.

Menzel JG, Wunderer H, Stavenga, DG (1991). Functional–morphology of the divided compound eye of the honeybee drone (*Apis-mellifera*). *Tissue Cell* 23: 525–535.

Michener CD (1974). *The Social Behavior of the Bees: A Comparative Study*. Harvard University Press, Cambridge, MA.

Michener CD (2007). *The Bees of the World*, 2nd ed. Johns Hopkins University Press, Baltimore, MD.

Milinski M, Croy I, Hummel T, Boehm T (2013). Major histocompatibility complex peptide ligands as olfactory cues in human body odour assessment. *Proc R Soc B* 280: 20122889.

Milinski M, Wedekind C (2001). Evidence for MHC-correlated perfume preferences in humans. *Behav Ecol* 12: 140–149.

Minister of Agriculture, Forestry and Fisheries (2013). Agricultural Pests Act 1983 (ACT No 36 of 1983): Control measures relating to honey-bees. *Government Gazette* 37015: 4–12.

Monceau K, Bonnard O, Moreau J, Thiery D (2014). Spatial distribution of *Vespa velutina* individuals hunting at domestic honeybee hives: heterogeneity at a local scale. *Insect Sci* 21: 765–774.

Moore D, Angel JE, Cheeseman IM, Fahrbach SE, Robinson GE (1998). Timekeeping in the honey bee colony: Integration of circadian rhythms and division of labor. *Behav Ecol Sociobiol* 43: 147–160.

Moritz RFA (1991a). Kin recognition in honeybees: Experimental artefact or biological reality? In Goodman LJ, Fisher RC (eds.), *The Behaviour and Physiology of Bees*. CAB International, Wallingford, UK, pp. 48–59.

Moritz RFA (1991b). The limitations of biometric control on pure race breeding in *Apis mellifera*. *J Apic Res* 30: 54–59.

Moritz RFA, de Miranda J, Fries I, Le Conte Y, Neumann P, Paxton RJ (2010). Research strategies to improve honeybee health in Europe. *Apidologie* 41: 227–242.

Moritz RFA, Erler S (2016). Lost colonies found in a data mine: Global honey trade but not pests or pesticides as a major cause of regional honeybee colony declines. *Agric Ecosyst Environ* 216: 44–50.

Moritz RFA, Härtel S, Neumann P (2005). Global invasions of the western honeybee (*Apis mellifera*) and the consequences for biodiversity. *Ecosci* 12: 289–301.

Moritz RFA, Lattorff HMG, Crous KL, Hepburn RH (2011). Social parasitism of queens and workers in the Cape honeybee (*Apis mellifera capensis*). *Behav Ecol Sociobiol* 65: 735–740.

Moritz RFA, Lattorff HMG, Neumann P, Kraus FB, Radloff SE, Hepburn HR (2005). Rare royal families in honeybees, *Apis mellifera*. *Naturwissenschaften* 92: 488–491.

Moritz RFA, Southwick EE (1991). *Bees as Superorganisms—An Evolutionary Reality*. Springer-Verlag, Berlin.

Mota T, Giurfa M (2010). Multiple reversal olfactory learning in honey bees. *Frontiers Behav Neurosci* 4: 48.

Mousseau TA, Roff DA (1987). Natural selection and the heritability of fitness components. *Heredity* 59: 181–197.

Nagari M, Bloch G (2012). The involvement of the antennae in mediating the brood influence on circadian rhythms in "nurse" honey bee (*Apis mellifera*) workers. *J Insect Physiol* 58: 1096–1103.

Naumann K (1992). *The Production and Intra-nest Transmission of Honey Bee (*Apis mellifera *L.) Queen Mandibular Gland Pheromone*. PhD thesis, Simon Fraser University, Vancouver, British Columbia, Canada.

Naumann K, Winston ML, Slessor KN, Prestwich GD, Webster FX (1991). Production and transmission of honey bee queen (*Apis mellifera*) queen mandibular pheromone. *Behav Ecol Sociobiol* 29: 321–332.

Nazzi F (2016). The hexagonal shape of the honeycomb cells depends on the construction behavior of bees. *Sci Rep* 6: 28341. doi:10.1038/srep28341

Nehring V, Evison SEF, Santorelli LA, d'Ettorre P, Hughes WHO (2011). Kin-informative recognition cues in ants. *Proc Biol Sci* 278: 942–1948.

Neumann P, Moritz RFA (2000). Testing genetic variance hypotheses for the evolution of polyandry in the honeybee (*Apis mellifera* L.). *Insect Soc* 47: 271–279.

Neumann P, Moritz RFA (2002). The Cape honeybee phenomenon: The evolution of a social parasite in real time? *Behav Ecol Sociobiol* 52: 271–281.

Neumann P, Moritz RFA, Mautz D (2000). Colony evaluation is not affected by drifting of drone and worker honeybees (*Apis mellifera* L.) at a performance testing apiary. *Apidologie* 31: 67–79.

Neumann P, Naef J, Crailsheim K, Crewe RM, Pirk CWW (2015). Hit-and-run trophallaxis of small hive beetles. *Ecol Evol* 5: 5479–5486.

Neumann P, Pirk CWW, Hepburn HR, Solbrig AJ, Ratnieks FLW, Elzen PJ, Baxter JR (2001). Social encapsulation of beetle parasites by Cape honey bee colonies (*Apis mellifera capensis* Esch.). *Naturwissenschaften* 88: 214–216.

Neumann P, van Praagh JP, Moritz RFA, Dustmann JH (1999). Testing reliability of a potential island mating apiary using DNA microsatellites. *Apidologie* 30: 257–276.

Nogueira-Neto P (1964). The spread of a fierce African bee in Brazil. *Bee World* 45: 119–121.

Nowak MA, Tarnita CE, Wilson EO (2010). The evolution of eusociality. *Nature* 466: 1057–1062.

Ogden J, Pardo C, Tchalpa A (1998). Method development for the analysis of bee cuticular hydrocarbons and esters by capillary gas chromatography. *J Chromatographr Sci* 36: 287–292.

Oi C, van Zweden JS, Oliveira RC, Van Oystaeyen A, Nascimento FS, Wenseleers T (2015). The origin and evolution of social insect queen pheromones: Novel hypotheses and outstanding problems. *BioEssays* 37: 808–821.

O'Keefe KJ, Schwarz MP (1990). Pheromones are implicated in reproductive differentiation in a primitively social bee. *Naturwissenschaften* 77: 83–86.

Oldroyd BP, Fewell JH (2007). Genetic diversity promotes homeostasis in insect colonies. *Trends Ecol Evol* 22: 408–413.

Oldroyd BP, Nanork P (2009). Conservation of Asian honey bees. *Apidologie* 40: 296–312.

Oldroyd BP, Pratt SC (2015). Comb Architecture of the Eusocial Bees Arises from Simple Rules Used During Cell Building. *Adv Insect Physiol* 49: 101–121.

Onions GW (1909). Bees: Workers and queens from eggs laid by workers. *Agric J Cape of Good Hope* 35: 527.

Onions GW (1912). South African "fertile-worker bees." *S Afr Agric J* 1: 720–728.

Oxley PR, Oldroyd BP (2010). The genetic architecture of honeybee breeding. *Adv Insect Physiol* 39: 83–118.

Page P, Lin Z, Buawangpong N, Zheng H, Hu F, Neumann P, Chantawannakul P, Dietemann V (2016). Social apoptosis in honey bee superorganisms. *Sci Rep* 6: 27210.

Page RE (1984). The evolution of multiple mating behavior by honey bee queens *Apis mellifera* L. *Genetics* 96: 263–273.

Page RE (2013). *The Spirit of the Hive: The Mechanisms of Social Evolution*. Harvard University Press, Cambridge, MA.

Page RE, Waddington KD, Hunt GJ, Fondrk MK (1995). Genetic determinants of honey bee foraging behavior. *Anim Behav* 50: 1617–1625.

Pahl M, Zhu H, Pix W, Tautz J, Zhang S (2007). Circadian timed episodic-like memory—A bee knows what to do when, and also where. *J Exp Biol* 210: 3559–3567.

Palmer KA, Oldroyd BP (2000). Evolution of multiple mating in the genus *Apis*. *Apidologie* 31: 235–248.

Paxton RJ (2005). Male mating behavior and mating systems of bees: An overview. *Apidologie* 36: 145–156.

Peeters CP, Monnin T, Malosse C (1999). Cuticular hydrocarbons correlated with reproductive status in a queenless ant. *Proc Roy Soc B Biol Sci* 266: 1323–1327.

Peiren N, de Graaf DC, Vanrobaeys F, Danneels EL, Devreese B, Van Beeumen J, Jacobs FJ (2008). Proteomic analysis of the honey bee worker venom gland focusing on the mechanisms of protection against tissue damage. *Toxicon* 52: 72–83.

Pereboom JJM, Velthuis HHW, Duchateau MJ (2003). The organisation of larval feeding in bumblebees and its significance to caste differentiation. *Insectes Soc* 50: 127–133.

Pflugfelder J, Koeniger N (2003). Fight between virgin queens (*Apis mellifera*) is initiated by contact to the dorsal abdominal surface. *Apidologie* 34: 249–256.

Pinto MA, Henriques D, Chavez-Galarza J, Kryger P, Garnery L, van der Zee R, Dahle B, Soland-Reckeweg G, de la Rua P, Dall'Olio R, et al. (2014). Genetic integrity of the dark European honey bee (*Apis mellifera mellifera*) from protected populations: A genome-wide assessment using SNPs and mtDNA sequence data. *J Apic Res* 53: 269–278.

Pinto MA, Rubink WL, Coulson RN, Patton JC, Johnston JS (2004). Temporal pattern of Africanization in a feral honeybee population from Texas inferred from mitochondrial DNA. *Evolution* 58: 1047–1055.

Pinto MA, Rubink WL, Patton JC, Coulson RN, Johnston JS (2005). Africanization in the United States: Replacement of feral European honeybees (*Apis mellifera* L.) by an African hybrid swarm. *Genetics* 170: 1653–1665.

Pirk CWW, Crewe RM, Moritz RFA (2017). Risks and benefits of the biological interface between managed and wild bee pollinators. *Funct Ecol* 31: 47–55. doi:10.1111/1365-2435.12768

Pirk CWW, Hepburn HR, Radloff SE, Tautz J (2004). Honey bee combs: Construction through a liquid equilibrium process? *Naturwissenschaften* 91: 350–353.

Pirk CWW, Human H, Crewe RM, VanEngelsdorp D (2014). A survey of managed honey bee colony losses in the Republic of South Africa—2009 to 2011. *J Apic Res* 53: 35–42.

Pirk CWW, Neumann P, Hepburn HR (2002). Egg laying and egg removal by workers are positively correlated in queenright Cape honey bee colonies (*Apis mellifera* capensis). *Apidologie* 33: 203–211.

Poinar GO, Danforth BN (2006). A fossil bee from Early Cretaceous Burmese amber. *Science* 314: 614–615.

Rangel J, Giresi M, Pinto MA, Baum KA, Rubink WL, Coulson RN, Johnston JS (2016). Africanization of a feral honey bee (*Apis mellifera*) population in south Texas: Does a decade make a difference? *Ecol Evol* 6(7): 2158–2169. doi:10.1002/ece3.1974

Rangel J, Griffin SR, Seeley TD (2010). An oligarchy of nest-site scouts triggers a honeybee swarm's departure from the hive. *Behav Ecol Sociobiol* 64: 979–987.

Rangel J, Mattila HR, Seeley TD (2009). No intracolonial nepotism during colony fissioning in honey bees. *Proc Biol Sci* 276: 3895–3900.

Rangel J, Reeve HK, Seeley TD (2013). Optimal colony fissioning in social insects: Testing an inclusive fitness model with honey bees. *Insect Soc* 60: 445–452.

Rangel J, Seeley TD (2008). The signals initiating the mass exodus of a honeybee swarm from its nest. *Anim Behav* 76: 1943–1952.

Ratnieks FLW (1991). The evolution of genetic odor-cue diversity in social Hymenoptera. *Am Nat* 137: 202–226.

Ratnieks FLW, Visscher PK (1989). Worker policing in the honey bee. *Nature* 342: 796–797.

Rauschmayer F (1928). Das Verfliegen der Bienen und die optische Orientierung am Bienenstand. *Arch Bienenkunde* 9: 249–322.

Rembold H, Hanser G (1964). Über den Weiselzellenfuttersaft der Honigbiene, VIII. Nachweis des determinierenden Prinzips im Futtersaft der Königinnenlarven. *Hoppe-Seyler Z Physiol Chem* 339: 251–254.

Rembold H, Lackner B (1981). Rearing of honey bee larvae in vitro: Effect of yeast extract on queen differentiation. *J Apic Res* 20: 165–171.

Rembold H, Lackner B, Geistbeck I (1974). The chemical basis of honey bee, *Apis mellifera*, caste formation: Partial purification of queen bee determinator from royal jelly. *J Insect Physiol* 20: 307–314.

Rice A, Schneider I, Barnett S, Cowan S (2016). *NAQS Asian honey bee floral surveillance manual.* Northern Australia Quarantine Service, Department of Agriculture and Water Resources, Canberra, August. CC BY 3.0.

Richards J, Carr-Markell M, Hefetz A, Grozinger CM, Mattila HR (2015). Queen-produced volatiles change dynamically during reproductive swarming and are associated with changes in honey bee (*Apis mellifera*) worker behavior. *Apidologie* 46: 679–690.

Rinderer TE, Hellmich RL, Danka RG, Collins AM (1985). Male reproductive parasitism: A factor in the Africanization of European honeybee populations, *Science* 228: 1119–1121.

Rinderer TE, Stelzer JA, Oldroyd BP, Buco SM, Rubink WL (1991). Hybridization between European and Africanized honey bees in the neotropical Yucatan Peninsula. *Science* 253: 309–311.

Roberts JMK, Anderson DL, Tay WT (2015). Multiple host shifts by the emerging honeybee parasite, *Varroa jacobsoni. Mol Ecol* 24: 2379–2391.

Robinson GE, Page RE, Huang Z-Y (1994). Temporal polyethism in social insects is a developmental process. *Anim Behav* 48: 467–469.

Rodriguez-Zas SL, Southey BR, Shemesh Y, Rubin EB, Cohen M, Robinson GE, Bloch G (2012). Microarray analysis of natural socially regulated plasticity in circadian rhythms of honey bees. *J Biol Rhythms* 27: 12–24.

Romiguier J, Cameron SA, Woodard SH, Fischman BJ, Keller L, Praz, CJ (2016). Phylogenomics controlling for base compositional bias reveals a single origin of eusociality in corbiculate bees. *Mol Biol Evol* 33: 670–678.

Rosenkranz P (1999). Honeybee (*Apis mellifera* L.) tolerance to *Varroa jacobsoni* Oud. in South America. *Apidologie* 30: 159–172.

Rosenkranz P, Aumeier P, Ziegelmann B (2010). Biology and control of *Varroa destructor*. *J Invertebrate Pathol* 103: S96–S119.

Rothenbuhler WC, Thompson VC (1956). Resistance to American foulbrood in honey bees: 1. Differential survival of larvae of different genetic lines. *J Econ Entomol* 49: 470–475.

Roubik DW (1978). Competitive interactions between neotropical pollinators and Africanized honey bees. *Science* 201: 1030–1032.

Roubik DW (1980). Foraging behavior of competing Africanized honeybees and stingless bees. *Ecology* 61: 836–845.

Roubik DW (1981) Comparative foraging behavior of apis-mellifera and trigona-corvina (hymenoptera, apidae) on baltimora-recta (compositae). *Revista Biol Trop* 29: 177–183.

Roubik DW (1982). Obligate necrophagy in a social bee. *Science* 217: 1059–1060.

Roubik DW (1983). Experimental community studies—Time-series tests of competition between African and neotropical bees. *Ecology* 64: 971–978.

Roubik DW (2000). Pollination system stability in tropical America. *Conserv Biol* 14: 1235–1236.

Roubik DW (2001). Ups and downs in pollinator populations: When is there a decline? *Conserv Ecol* 5: 2.

Roubik DW, Buchmann SL (1984). Nectar selection by *Melipona* and *Apis mellifera* (Hymenoptera, Apidae) and the ecology of nectar intake by bee colonies in a tropical forest. *Oecologia* 61: 1–10.

Roubik DW, Reyes F (1984). African honey bees have not brought Acarine mite infestations to Panama. *Am Bee J* 124: 665–667.

Roubik DW, Villanueva-Gutierrez R (2009). Invasive Africanized honey bee impact on native solitary bees: A pollen resource and trap nest analysis. *Biol J Linnean Soc* 98: 152–160.

Roubik DW, Wolda H (2001). Do competing honey bees matter? Dynamics and abundance of native bees before and after honey bee invasion. *Popul Ecol* 43: 53–62.

Ruttner F (1988). *Biogeography and Taxonomy of Honeybees*. Springer Verlag, Berlin.

Ruttner F, Maul V (1983). Experimental analysis of reproductive interspecies isolation of *Apis mellifera* L. and *Apis cerana* F. *Apidologie* 14: 309–327.

Sakagami SF (1958). The false-queen: Fourth adjustive response in dequeened honey bee colonies. *Behaviour* 13: 280–296.

Sakagami SF (1959). Interspecific relations between Japanese and European honeybees. *J Anim Ecol* 28: 51–68.

Sakagami SF, Akahira Y (1960). Studies on the Japanese honey bee, *Apis cerana cerana* Fabricius: VIII. Two opposing adaptations in the post-stinging behavior of honey bees. *Evolution* 14: 29–40.

Sammataro D, Cicero JM (2010). Functional morphology of the honey stomach wall of the European honey bee (Hymenoptera: Apidae). *Ann Entomol Soc Am* 103: 979–987.

Sandoz JC (2006). Odour-evoked responses to queen pheromone components and to plant odours using optical imaging in the antennal lobe of the honey bee drone *Apis mellifera* L. *J Exp Biol* 209: 3587–3598.

Sanguinetti A, Singer RB (2014). Invasive bees promote high reproductive success in Andean orchids. *Biol Conserv* 175: 10–20.

Santos LG, Alves MLTMF, Message D, Pinto FA, Silva MVGB, Teixeira EW (2014). Honey bee health in Apiaries in the Vale do Paraiba, Sao Paulo State, southeastern Brazil. *Sociobiology* 61: 307–312.

Sakai T, Okada I (1973). The present beekeeping in Japan. *Glean Bee Cult* 101: 356–357.

Sauer S, Herrmann E, Kaiser W (2004). Sleep deprivation in honeybees. *J Sleep Res* 13: 145–152.

Sauer S, Kinkelin M, Herrmann E, Kaiser W (2003). The dynamics of sleep-like behaviour in honeybees. *J Comp Physiol A* 189: 599–607.

Schlegel T, Visscher PK, Seeley TD (2012). Beeping and piping: Characterization of two mechano-acoustic signals used by honey bees in swarming. *Naturwissenschaften* 99: 1067–1071.

Schlüns H, Schlüns EA, van Praagh J, Moritz RFA (2003). Sperm numbers in drone honeybees (Apis mellifera) depend on body size. *Apidologie* 34: 577–584.

Schlüns H, Koeniger G, Koeniger N, Moritz RFA (2004). Sperm utilization pattern in the honeybee (*Apis mellifera*). *Behav Ecol Sociobiol* 56: 458–463.

Schmid-Hempel P (1998). *Parasites in Social Insect*. Princeton University Press, Princeton, NJ.

Schmidt JO, Thoenes SC (1990). Honey bee (Hymenoptera, Apidae) preferences among artificial nest cavities. *Ann Entomol Soc Am* 83: 271–274.

Schneider SS (1990). Nest characteristics and recruitment behavior of absconding colonies of the African honey bee, *Apis mellifera scutellata*, in Africa. *J Insect Behav* 3: 225–240.

Schneider SS, Blyther R (1988). The habitat and nesting biology of the African honey bee *Apis mellifera scutellata* in the Okavango River Delta, Botswana, Africa. *Insect Soc* 35: 167–181.

Schneider SS, Hoffman GD, Smith DR (2004). The African honey bee: Factors contributing to a successful biological invasion. *Annu Rev Entomol* 49: 351–376.

Schulte C, Theilenberg E, Müller-Borg M, Gempe T, Beye M (2014). Highly efficient integration and expression of piggyBac-derived cassettes in the honey bee (*Apis mellifera*). *Proc Natl Acad Sci USA* 111: 9003–9008.

Schumacher MJ, Schmidt JO, Egen NB (1989). Lethality of "killer bee" stings. *Nature* 337: 413.

Schumacher MJ, Schmidt JO, Egen NB, Lowry JE (1990). Quantity analysis and lethality of European and Africanized honey bee venoms. *Am J Trop Med Hyg* 43: 79–86.

Schumacher MJ, Tveten MS, Egen NB (1994). Rate and quantity of delivery of venom from honey bee stings. *J Allergy Clin Immunol* 93: 831–835.

Schwartz MP, Tierney SM, Rehan SM, Chenoweth LB, Cooper SJB (2011). The evolution of eusociality in allodapine bees: Workers began by waiting. *Biol Lett* 7: 277–2806.

Seeley TD (1977). Measurement of nest cavity volume by the honeybee (*Apis mellifera*). *Behav Ecol Sociobiol* 2: 201–227.

Seeley TD (1985). *Honey Bee Ecology*. Princeton University Press, Princeton, NJ.

Seeley TD (1996). *The Wisdom of the Hive: The Social Physiology of Honey Bee Colonies*. Harvard University Press, Cambridge, MA.

Seeley TD (2010). *Honeybee Democracy*. Princeton University Press, Princeton, NJ.

Seeley TD, Tarpy DR (2007). Queen promiscuity lowers disease within honeybee colonies. *Proc Roy Soc—Biol Sci* 274: 67–72.

Seeley TD, Smith ML (2015). Crowding honeybee colonies in apiaries can increase their vulnerability to the deadly ectoparasite *Varroa destructor. Apidologie* 46: 716–727.

Seidl R (1982). *Die Sehfelder und Ommatidien-Divergenzwinkel von Arbeiterin, Königin und Drohne der Honigbiene* (Apis mellifica). PhD thesis. Technical University, Darmstadt, Germany.

Sendova-Franks A, Franks NR (1993). Task allocation in ant colonies within variable environments (A study of temporal polyethism: Experimental). *Bull Math Biol* 55: 75–96.

Shemesh Y, Cohen M, Bloch G (2007). Natural plasticity in circadian rhythms is mediated by reorganization in the molecular clockwork in honey bees. *FASEB J* 21: 2304–2311.

Shemesh Y, Eban-Rothschild A, Cohen M, Bloch G (2010). Molecular dynamics and social regulation of context-dependent plasticity in the circadian clockwork of the honey bee. *J Neurosci* 30: 12517–12525.

Simone-Finstrom M, Spivak M (2010). Propolis and bee health: The natural history and significance of resin use by honey bees. *Apidologie* 41: 295–311.

Snodgrass RE (1956). *Anatomy of the Honey Bee*. Comstock, New York.

Soland-Reckeweg G, Heckel G, Neumann P, Fluri P, Excoffier L (2009). Gene flow in admixed populations and implications for the conservation of the western honeybee, *Apis mellifera. J Insect Conserv* 13: 317–328.

Soro A, Quezada-Euan JJG, Theodorou P, Moritz RFA, Paxton RJ (2017). The population genetics of two orchid bees suggests high dispersal, low diploid male production and only an effect of island isolation in lowering genetic diversity. *Conserv Genet* 18(3): 607–619. doi:10.1007/s10592-016-0912-8

Southwick EE (1985). Allometric relations, metabolism and heat conductance in clusters of honey bees at cool temperatures. *J Comp Physiol B* 156: 143–149.

Spivak M (1992). The relative success of Africanized and European honeybees over a range of life zones in Costa Rica. *J Appl Ecol* 29: 150–162.

Stanley-Samuelson DW, Jurenka RA, Cripps C, Blomquist GJ, de Renobales M (1988). Fatty acids in insects: Composition, metabolism, and biological significance. *Arch Insect Biochem Physiol* 9: 1–33.

Steinhobel F (1977). Swarming on the aloes and its utilisation for making increase. In Fletcher DJC (ed.), Apimondia Symposium: *African Bees: Taxonomy, Biology and Economic Use.* Apimondia, Pretoria, South Africa, pp. 152–156.

Stewart JB, Chinnery PF (2015). The dynamics of mitochondrial DNA heteroplasmy: Implications for human health and disease. *Nature Rev Genet* 16: 530–542.

Strauss U, Dietemann V, Human H, Crewe RM, Pirk CWW (2016). Resistance rather than tolerance explains survival of savannah honey bees (*Apis mellifera scutellata*) to infestation by the parasitic mite *Varroa destructor. Parasitology* 143: 374–387.

Strauss U, Human H, Gauthier L, Crewe RM, Dietemann V, Pirk CWW (2013). Seasonal prevalence of pathogens and parasites in the savannah honeybee (*Apis mellifera scutellata*). *J Invertebrate Pathol* 114: 45–52.

Streinzer M, Brockmann A, Nagaraja N, Spaethe J (2013). Sex and caste-specific variation in compound eye morphology of five honeybee species. *PLoS One* 8(2): e57702. doi:10.1371/journal.pone.0057702

Stuart RJ, Herbers JM (2000). Nest mate recognition in ants with complex colonies: within- and between-population variation. *Behav Ecol* 11: 676–685.

Swammerdam J (1737). Verhandeling van de Byen. In Boerhave H (ed.), *Bybel der natuure of historie der insecten*. Facsimile reprint (1980), De Banier/De Groot, Utrecht/Goudriaan, the Netherlands, pp. 367–550.

Tanner DA, Visscher PK (2010). Adaptation or constraint? Reference-dependent scatter in honey bee dances. *Behav Ecol Sociobiol* 64: 1081–1086.

Taylor OR (1977). Past and possible future spread of Africanized honeybees in Americas. *Bee World* 58: 19–30.

Teixeira EW, Chen Y, Message D, Pettis J, Evans JD (2008). Virus infections in Brazilian honey bees. *J Invertebrate Pathol* 99: 117–119.

Teixeira LV, Waterhouse JM, Marques MD (2011). Respiratory rhythms in stingless bee workers: Circadian and ultradian components throughout adult development. *J Comp Physiol A* 197: 361–372.

The Swarm, directed by Irwin Allen. Warner Brothers, Burbank, CA (1978).

Thorp RW, Briggs DL (1980). Bees collecting pollen from other bees Hymenoptera Apoidea. *J Kansas Entomol Soc* 53: 166–170.

Tofts C (1993). Algorithms for task allocation in ants (a study of temporal polyethism: Theory). *Bull Math Biol* 55: 891–918.

Tofts C, Franks NR (1992). Doing the right thing: Ants, honey bees and naked mole rats. *Trends Ecol Evol* 7: 346–349.

Toma DP, Bloch G, Moore D, Robinson GE (2000). Changes in period mRNA levels in the brain and division of labor in honey bee colonies. *Proc Natl Acad Sci USA* 97: 6914–6919.

Underwood RM, Lewis MJ, Hare JF (2004). Reduced worker relatedness does not affect cooperation in honey bee colonies. *Canadian J Zool* 82: 1542–1545.

van Engelsdorp D, Evans JD, Saegerman C, Mullin C, Haubruge E, Nguyen BK, Frazier M, Frazier J, Cox-Foster D, Chen Y, et al. (2009). Colony collapse disorder: A descriptive study. *PLoS One* 4(8): e6481.

van Engelsdorp D, Underwood R, Caron D, Hayes J Jr (2007). An estimate of managed colony losses in the winter of 2006–2007: A report commissioned by the Apiary Inspectors of America. *Am Bee J* 147: 599–603.

van Nest BN, Moore D (2012). Energetically optimal foraging strategy is emergent property of time-keeping behavior in honey bees. *Behav Ecol* 23: 649–658.

Van Oystaeyen A, Oliveira RC, Holman L, van Zweden JS, Romero C, Oi CA, d'Etorre P, Khalesi M, Billen J, Wackers F, et al. (2014). Conserved class of queen pheromones stops social insect workers from reproducing. *Science* 343: 287–290.

van Wilgenburg E, Driessen G, Beukeboom LW (2006). Single locus complementary sex determination in Hymenoptera: An "unintelligent" design? *Frontiers Zool* 3: 1.

Vogt H (1911). *Geometrie und Ökonomie der Bienenzelle*. Trewendt und Granier, Breslau, Poland.

von Buttel-Reepen HB (1900). *Sind die Bienen Reflexmaschinen?* Verlag Arthur Georgi, Leipzig, Germany.

von Frisch K (1940). Die Tänze und das Zeitgedächtnis im Widerspruch. *Naturwissenschaften* 28: 65–69.

von Frisch K (1965). *Tanzsprache und Orientierung der Bienen*. Springer-Verlag, Berlin.

von Frisch K (1967). *The Dance Language and Orientation of Bees*. Harvard University Press, Cambridge, MA.

Wagner AE, Van Nest BN, Hobbs CN, Moore D (2013). Persistence, reticence and the management of multiple time memories by forager honey bee. *J Exp Biol* 216: 1131–1140.

Wallberg A, Han F, Wellhagen G, Dahle B, Kawata M, Haddad N, Simões ZLP, Allsopp MH, Kandemir I, De la Rúa P, et al. (2014). A worldwide survey of genome sequence variation provides insight into the evolutionary history of the honeybee *Apis mellifera*. *Nat Genet* 46: 1081–1088.

Wanner KW, Nichols AS, Walden KKO, Brockmann A, Luetje CW, et al. (2007). A honey bee odorant receptor for the queen substance 9-oxo-2-decenoic acid. *Proc Natl Acad Sci USA* 104: 14383–14388.

Warrant E, Porombka T, Kirchner WH (1996). Neural image enhancement allows honey bees to see at night. *Proc Biol Sci* 263: 1521–1526.

Wattanachaiyingcharoen W, Oldroyd BP, Wongsiri S, Palmer K, Paar R (2003). A scientific note on the mating frequency of *Apis dorsata*. *Apidologie* 34: 85–86.

Wei CA, Dyer FC (2009). Investing in learning: Why do honey bees, *Apis mellifera*, vary the durations of learning flights? *Anim Behav* 77: 1165–1177.

Wenner AM (2002). The elusive honey bee dance "language" hypothesis. *J Insect Behav* 15: 859–878.

Wenner AM, Wells PH (1990). *Anatomy of a Controversy: The Question of a "Language" Among Bees*. Columbia University Press, New York.

Wharton KE, Dyer FC, Huang ZY, Getty T (2007). The honeybee queen influences the regulation of colony drone production. *Behav Ecol* 18: 1092–1099.

Wheeler WM (1911). The ant-colony as an organism. *J Morphol* 22: 307–325.

Whitfield CW, Behura SK, Berlocher SH, Clark AG, Johnston JS, Sheppard WS, Smith DR, Suarez AV, Weaver D, Tsutsui ND (2006). Thrice out of Africa: Ancient and recent expansions of the honey bee, *Apis mellifera*. *Science* 314: 642–645.

Whiting PW (1943). Multiple alleles in complementary sex determination in *Habrobracon*. *Genetics* 28: 365–382.

Williams JL (1987). Wind-directed pheromone trap for drone honey-bees (Hymenoptera, Apidae). *J Econ Entomol* 80: 532–536.

Winston ML (1987). *The Biology of the Honey Bee*. Harvard University Press, Cambridge, MA.

Woyke J (1965). Study on the comparative viability of diploid and haploid larval drone honeybees. *J Apic Res* 4: 12–16.

Woyke J (1969). A method of rearing diploid drones in a honey bee colony. *J Apic Res* 8: 65–74.

Woyke J (1980). Genetic background of sexuality in the diploid drones honey bee. *J Apic Res* 19: 89–95.

Wright GA, Nicolson SW, Shafir S (2018). Nutritional physiology and ecology of honey bees. *Annu Rev Entomol* 63: 327–344.

Yang GH (2005). Harm of introducing the western honeybee *Apis mellifera* L. to the Chinese honeybee *Apis cerana* F. and its ecological impact. *Acta Entomol Sin* 48: 401–406.

Yang M-X, Tan K, Radloff SE, Hepburn HR (2011). Interspecific interactions among Asian honey bees. In Hepburn HR, Radloff SE (eds.), *Honey Bees of Asia*. Springer-Verlag, Berlin, pp. 445–472.

Yoder JA, Jajack AJ, Rosselot AE, Smith TJ, Yerke MC, Sammataro D (2013). Fungicide contamination reduces beneficial fungi in bee bread based on an area-wide field study in honey bee, *Apis mellifera*, colonies. *J Toxicol Environ Health Part A* 76: 587–600.

{ INDEX }

The manufacturer's authorised representative in the EU for product safety is Oxford University Press España S.A. of El Parque Empresarial San Fernando de Henares, Avenida de Castilla, 2 - 28830 Madrid (www.oup.es/en or product.safety@oup.com). OUP España S.A. also acts as importer into Spain of products made by the manufacturer.
Printed and bound by CPI Group (UK) Ltd, Croydon, CR0 4YY
14/09/2026
02216310-0001